Entreprenuer SKills

Taken from This Lesso[illegible]
about How to Read Electronics
& Circuits Diagrams.

Adjective describing
The story Indicates an
Opinion is not Included
in Objective Summary

Linking Verb Connect or link
The Subject to a word or wordgroups
object That describes or Identify
The Subject.

Possessive Pronouns are used to Replace
a noun Phrase that Starts with the Possessive
~~Ad~~ Adjectives

How to Read Electronic Circuit Diagrams

2nd Edition

Robert M. Brown, Paul Lawrence,
and James A. Whitson

TAB Books
Division of McGraw-Hill
New York San Francisco Washington, D.C. Auckland Bogotá
Caracas Lisbon London Madrid Mexico City Milan
Montreal New Delhi San Juan Singapore
Sydney Tokyo Toronto

pbk 14 15 16 17 18 19 20 FGR/FGR 9 9 8
hc 2 3 4 5 6 7 8 9 10 11 FGR/FGR 8 9

Library of Congress Cataloging-in-Publication Data

Brown, Robert Michael, 1943–
How to read electronic circuit diagrams / by Robert M. Brown, Paul Lawrence, and James Whitson.—2nd ed.
p. cm.
Includes index.
ISBN 0-8306-0480-4 ISBN 0-8306-2880-0 (pbk.)
1. Electronics—Charts, diagrams, etc. I. Lawrence, Paul, 1934– .
II. Whitson, James A. III. Title.
TK7866.B7 1987
621.381′3—dc19

87-29027
CIP

HT3
2880

OTHER BOOKS IN THE TAB HOBBY ELECTRONICS SERIES

This series of five newly updated and revised books provides an excellent blend of theory, skills, and projects that lead the novice gently into the exciting arena of electronics. The Series is also an extremely useful reference set for libraries and intermediate and advanced electronics hobbyists, as well as an ideal text.

The first book in the set, *Basic Electronics Course—2nd Edition*, consists of straight theory and lays a firm foundation upon which the Series then builds. Practical skills are developed in the second and third books. *How to Read Electronic Circuit Diagrams—2nd Edition* and *How to Test Almost Anything Electronic—2nd Edition*, which cover the two most important and fundamental skills necessary for successful electronics experimentation. The final two volumes in the Series, *44 Power Supplies for Your Electronic Projects* and *Beyond the Transistor: 133 Electronics Projects*, present useful hands-on projects that range from simple half-wave rectifiers to sophisticated semiconductor devices utilizing ICs.

Basic Electronics Course—*2nd Edition*

NORMAN H. CROWHURST

This book thoroughly explains the necessary fundamentals of electronics, such as electron flow, magnetic fields, resistance, voltage, and current.

How to Test Almost Everything Electronic—*2nd Edition*

JACK DARR and DELTON T. HORN

This book describes electronic tests and measurements—how to make them with all kinds of test equipment, and how to interpret the results. New sections cover logic probes and analyzers, using a frequency counter and capacitance meter, signal tracing a digital circuit, identifying unknown ICs, digital signal shaping, loading and power supply problems in digital circuits, and monitoring brief digital signals.

44 Power Supplies for Your Electronic Projects

ROBERT J. TRAISTER and JONATHAN L. MAYO

Electronics and computer hobbyists will not find a more practical book than this one. A quick, short, and thorough review of basic electronics is provided along with indispensable advice on laboratory techniques and how to locate and store components. The projects begin with simple circuits and progress to more complicated designs that include ICs and discrete components.

Beyond the Transistor: 133 Electronics Projects

RUFUS P. TURNER and BRINTON L. RUTHERFORD

Powerful integrated circuits—both digital and analog—are now readily available to electronic hobbyists and experimenters. And many of these ICs are used in the exciting projects described in this new book: a dual LED flasher, an audible continuity checker, a proximity detector, a siren, a pendulum clock, a metronome, and a music box, just to name a few. For the novice, there is information on soldering, wiring, breadboarding, and troubleshooting, as well as where to find electronics parts.

Contents

Acknowledgments vii

Introduction viii

1 Basic Components 1

Resistors • Capacitors • Coils and Chokes • Transformers • Batteries • Switches • Relays • Fuses • Circuit Breakers

2 Transducers, Indicating Components, and Miscellaneous Components 25

Crystals • Filters • Headphones • Lamps • Meters • Microphones • Phono Pickups • Speakers • Tape Recorder Heads • Heatsinks • Experimenter's Breadboard

3 Solid-State Devices 42

Transistors • Diodes • Integrated Circuits • Special Devices • Field-Effect Transistors (FETs) • IGFETs and MOSFETs • Unijunction Transistor • LEDs • LCDs • Photocells • Infrared Emitter and Detector • Thermistors

4 Vacuum Tubes 57

Diodes • Triodes • Tetrode • Pentode • Beam Power Tubes • Pentagrid Tubes • Regulator Tubes • Cathode-Ray Tubes

5 Interconnecting Devices 66

Printed-Circuit Boards • Wiring Harnesses • Multiple-Wire Connectors • Audio Connectors • Special Connectors • Special-Purpose Cable

6 Types of Diagrams 77

Schematic Diagrams • Block Diagrams • Layout Diagrams • Pictorial Diagrams • Mechanical Construction Diagrams • Drawing Electronic Diagrams

7 **Radio and TV Schematics** 91

The Heathkit Portable AM Radio, Model GR-1009 • The Heathkit 19″ Color TV with Remote Control, Model GR-1903

8 **Specialized Electronic Equipment** 151

The Heathkit Dual-Trace Oscilloscope, Model IO-4210 10-MHz

9 **Understanding Digital Circuits** 174

Number Systems • Boolean Algebra • Digital Gates • Reading Digital Schematics • Measuring Logic Levels

Appendix A Electronic Schematic Symbols 186

Appendix B Color Codes 195

Appendix C Wires Sizes 198

Appendix D Terms and Abbreviations 200

Glossary 203

Index 212

Acknowledgments

I would like to express my thanks to the following individuals and companies that have provided information for this book: F. Wiley Hunt and Jim Brow of Heathkit; Bill Porter and Carol Parcels of Hewlett Packard; David S. Gunzel of Radio Shack, a division of Tandy Corporation; Carol Cannon of Thermalloy, Inc.; and Jim Wallace of Supertex, Inc.

Introduction

Electronic circuit diagrams are the keys to understanding the functioning of electronic circuits and electronic equipment. Over the years, since the discovery of electricity and the development of electronics, there has arisen a series of conventions for denoting the individual electronic components and their connections in different electronic circuits. This book attempts to explain what those conventions are and how they are used.

Begin with the basic electronic schematic symbols and some simple circuits. These symbols are the basis for all further developments in electronics, and they must be mastered before going into more intricate areas, such as digital electronics. In most cases, both the electronic symbol and an actual photograph or drawing of the component are shown. In this fashion, the reader will easily be able to relate the abstract symbol with the real component that it represents. Gradually, more complete circuits are introduced, using more and more types of components. In this way, all different types of electronic components, from simple resistors to highly complex ICs (integrated circuits), are covered.

In this second edition, new material has been added concerning digital electronics. There has been an extraordinary growth in the development of this branch of electronics in the years since the first edition was published. Other areas of electronics—especially basic analog circuits—have changed, also. Today, it is common practice to use analog integrated circuits for applications that would have required a large number of discrete devices just a few years ago.

This book is for both the beginner in electronics and for the more experienced reader who might want to brush up on reading electronic circuit diagrams. In Chapters 7 and 8, the reader will find some examples of what superior electronic circuit diagrams should look like. These diagrams have been provided by the fine folks at Heathkit. Heathkits are manufactured by the Heath Company in

Benton Harbor, Michigan. This company provides some of the very best electronic products for the hobbyist and experimenter. If you have ever built a Heathkit electronic product, you know that their instructions, drawings, and schematics are all excellent. If you haven't, then you are in for a real treat!

1
Basic Components

The hundreds of parts that go into making a radio, hi-fi, or television set work are called *circuit components*. These include the resistors, capacitors, tubes, coils, etc., that are necessary in the operation of any electronic circuit. All components are identified by a symbol on the schematic diagram of the equipment, much like the symbols an architect uses to show the stairs, doors, and walls on the floor plan of a home.

The schematic symbol for most components simply tells you what that particular component is. It does not tell you what the electrical characteristics are, just as the symbol an architect uses for stairs does not tell you what kind of wood is used to make the stairs. To find the value and ratings of any component, simply refer to the parts list. However, some components do have their values either stamped on them or printed in the form of color-coded dots or bands. This is the case with most resistors as well as some capacitors and coils. This color code is standard in electronics in accordance with established Electronics Industries Associated (EIA) specifications. An explanation of what this code is and how it works is provided in Appendix B.

There are several ways in which a circuit component can be identified or described: 1) by means of a picture or diagram, 2) by a letter, and 3) by a schematic symbol. Components discussed in this chapter are identified by their standard letter designation and schematic symbol. Pictures or diagrams of all the many types of resistors, capacitors, etc., would fill volumes and be of little benefit here. In this book, pictures of the components are accompanied by their corresponding schematic symbols in the lower right corner of the photo.

RESISTORS

Resistors are found in almost every type of electronic circuit. Resistors are devices used to restrict electrical current flow and produce a voltage drop. The variety of material used to make resistors is much more limited than it is for capacitors, but, like capacitors, they are available in both *fixed* and *variable* types and some varieties which are *tapped*.

Many basic types of resistors are available. They can be made of carbon, cermet (a metal-glaze configuration), wirewound, or made with metal film, which is vacuum-deposited over a rectangular- or cylinder-shaped substrate. These materials are selected to provide a specific amount of resistance to current flow and also for their respective tolerance capabilities. Carbon resistors are available in many sizes, with tolerances as low as 1 percent. Wirewound, cermet, and film (also known as Film-met) resistors can be made with tolerances as low as .01 percent.

No matter what a resistor is composed of, if it is a fixed type, it can be identified by the schematic circuit symbol shown in Fig. 1-1(A). A variable resistor is shown in Fig. 1-1(B) and a tapped resistor in Fig. 1-1(C). As with capacitors, the symbol immediately reveals that the component in question is a resistor while simultaneously indicating whether it is a fixed, tapped, or variable type. But unlike capacitors, most fixed resistors are color-coded to show their exact resistance value and their inherent tolerance, defined as the ability to hold their designated value (refer to Appendix B for color-coding).

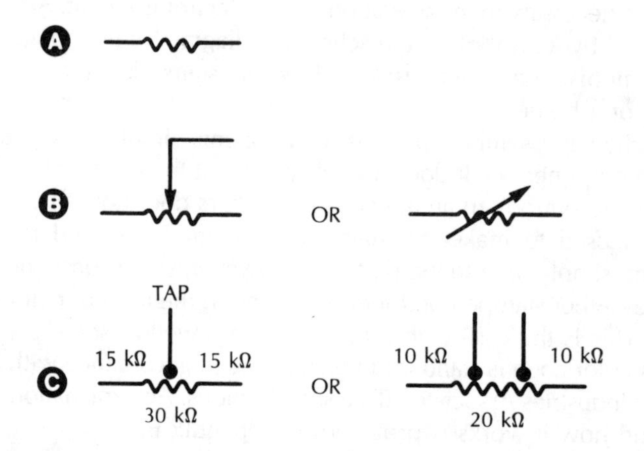

Fig. 1-1. Resistor schematic symbols (A) fixed, (B) variable, (C) tapped.

Another important consideration for any given resistor is its wattage rating, which can be chosen after circuit power requirements have been determined. If a resistor is used in a circuit where heavy current is flowing and its resistance value causes a large voltage drop, the resistor must be able to safely dissipate the resulting degree of heat that will be generated. Therefore, resistors are also rated by the amount of heat (watts) they can safely withstand. Physically, the larger the resistor the more heat it can dissipate, meaning that you can generally judge the wattage rating of many resistors simply by looking at them.

As indicated earlier, all components have a schematic reference letter/number, and for resistors it is the letter R. Suppose you see "R23" printed next to a resistor symbol. This means it is a resistor and the twenty-third one on the schematic. Again, you need this reference number to identify the component in the parts list. Resistor values also are printed near the symbol on most schematics.

The value or quantity of resistance is measured in terms of *ohms*. This term is given the Greek symbol Omega or Ω. So when you see "100Ω" next to a resistor, you know its value is 100 ohms. Resistor values range from fractions of an ohm into the thousands and millions of ohms. It would be ridiculous to try to print that many numbers on a schematic, so letters were adopted which represent one thousand and one million—the letter k, meaning *kilo* or one thousand, and the letter M, meaning *mega* or one million. Therefore, a 10,000-ohm resistor would be labeled 10K on a schematic while a 10-million ohm resistor would be labeled 10M or 10 Meg. The symbol for ohms (Ω) is often left off of the schematic because it is assumed.

Several types of resistors are shown in Fig. 1-2. A fixed resistor is a specific value in ohms. A tapped or adjustable resistor is a kind of semi-fixed resistor, usually wirewound with one or more slider arms, which are adjusted to a point along the length of the resistor to provide only a certain amount of total resistance. The tap is then tightened and left at this point. A tapped resistor differs from a variable resistor in that it is normally not changed in value once it is adjusted.

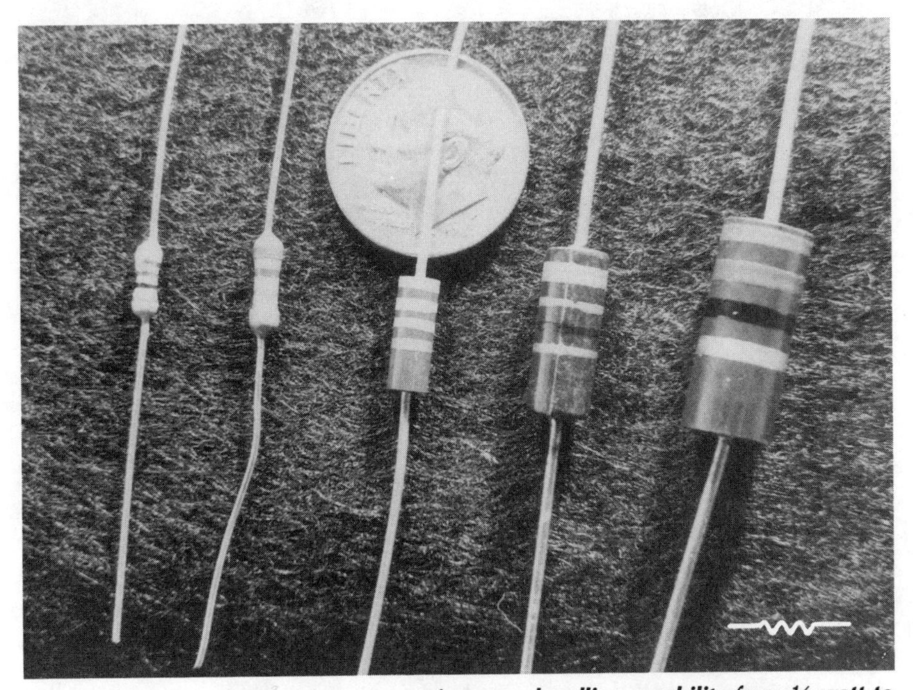

Fig. 1-2. (A) These fixed resistors range in power-handling capability from ⅛ watt to 2 watts. Schematic symbol shown in lower right corner.

(B) Small variable "trimmer resistors" and schematic symbol.

(C) 10-watt power resistor with symbol.

Fig. 1-3 shows how a fixed wirewound resistor is made.

A variable resistor, sometimes called a rheostat, potentiometer, or simply "pot," can be rotated or adjusted from zero to its full value as the operator desires. The volume control on your radio, hi-fi, and TV set are all examples of variable resistors or pots. See Fig. 1-4. Transistorized equipment often uses miniaturized versions of variable resistors called "trim pots." With the trend toward more miniaturization in all types of circuits, resistors (as well as other components) get smaller and smaller. Because the currents and voltages used in transistor and IC circuits are also small, the physical size of components is reduced considerably.

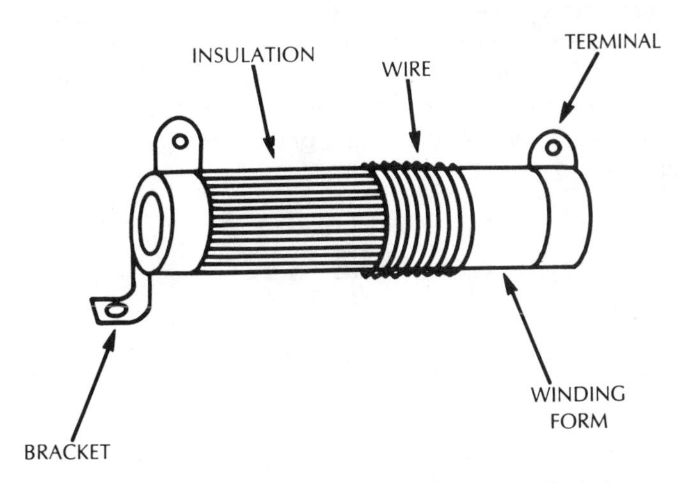

Fig. 1-3. The construction of a wirewound power resistor.

Fig. 1-4. This type of variable resistor (potentiometer) is often used as a volume control (Symbol in lower right corner).

CAPACITORS

A capacitor is used to "store" electricity. Basically, a capacitor is made of two plates or electrodes separated by some type of insulation such as air, mica, glass, or even oil. This insulation is known as the *dielectric* and it is one of the most important inherent properties of the capacitor. Although there are many types of available dielectrics, they can be generally grouped into three classes: air, solid, and electrolytic films. Mica, glass, and oil are examples of a solid dielectric.

Fundamentally, there are two types of capacitors: fixed or variable. Schematic symbols for both types are shown in Fig. 1-5. On most schematic diagrams, the letter C is used to designate a capacitor, regardless of its type. Neither the schematic symbol nor the letter will tell you what the dielectric is made of or the value of the capacitor. However, this information can often be determined from the capacitor's physical appearance or by referring to the equipment's parts list.

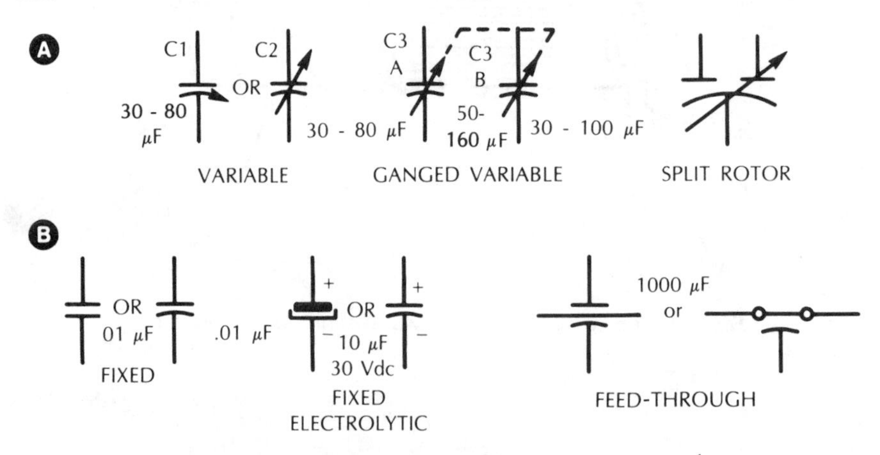

Fig. 1-5. Capacitor schematic symbols (A) variable, (B) fixed.

There are, though, other numbers and letters marked on the schematic along with the capacitor symbol that do provide some vital information. For example, if you look at a capacitor symbol on a typical schematic, it is generally identified by the letter C followed by a number printed close to the capacitor symbol. To elaborate, let us assume that C23 is listed next to the capacitor. This reveals that it is, in fact, a capacitor and it is number 23 in the circuit. The reference number is needed to identify the exact component in the parts list. In fact, every component is identified in the parts list with its own reference number.

The value or *capacitance* of a capacitor is also normally included on the schematic near the component. This may be expressed as simply .01 or .01μF. To briefly explain what this means, the letters μF (older schematics may use "mfd" or "mF") stand for microfarads and indicate the electrical size of the component. A microfarad is one one-millionth of a farad. If the value were labeled pF (on older schematics you may find μμF or even mmF), it would mean one one-millionth of a microfarad.

Note: The abbreviation μF (the Greek letter mu and the English letter F) are used on modern schematic diagrams. On older schematic diagrams (or schematics drawn by older electronics technicians and engineers!) you will often find the abbreviation "mfd" of "mF." These letters also stand for microfarad and you should not confuse them with *millifarad* (one one-thousandth of a farad). The term millifarad is almost never used; one one-thousandth of a farad is designated as 1000 μF or 0.001 F.

As shown by the arrows on the schematic symbols of Fig. 1-5(A), a variable capacitor is one that can be adjusted or turned. A simple variable capacitor would have two plates: one that is stationary and one that can be rotated. The plates of a variable capacitor normally have an air gap between them to act as the insulation or dielectric. See Fig. 1-6.

Fig. 1-6. (A) A single-section variable capacitor. (B) A three-section "ganged" variable capacitor. Both have similar symbols.

In application, however, a variable capacitor may have many fixed and movable plates. As the shaft (connected to the movable plates) is turned, the plates (known as *rotors*) move out of "mesh" with the fixed plates (known as *stators*). As a result, the capacitance—or value—of the capacitor is increased. When the fixed and movable plates are meshed together, the capacitance is at minimum. Therefore, a variable capacitor can be adjusted or varied from some minimum value to a maximum value.

For example, the tuning dial of a radio is generally connected indirectly to a variable capacitor, normally by a dial cord, which is in turn connected to the shaft of the capacitor. As you tune the dial of a radio, you are actually changing the capacity in the circuit, which causes a change in frequency. This type of capacitor frequently has two or more entire sections that are electrically separated but mechanically connected to a common tuning shaft. This type of capacitor is known as a *ganged variable*. The schematic symbol is shown in Fig. 1-5(A), and is pictured in Fig. 1-6.

Another type of variable capacitor is called a *trimmer*; some trimmers have an air dielectric while others use a solid dielectric such as mica, glass, or ceramic. Miniature airdielectric capacitors used for printed-circuit applications are shown in Fig. 1-7.

Fig. 1-7. Small trimmer capacitors for printed-circuit board use and symbol.

A *fixed* capacitor, as its name implies, has only one specific value or capacity. They come in hundreds of sizes and shapes with equally as many types of dielectric materials. Some of the more common fixed capacitors are the types shown in Fig. 1-8, as well as special types such as *feedthrough* and *bathtub* capacitors. Electrolytic capacitors are also fixed, but they are different electrically from other capacitors. Electrolytics use a specific type of dielectric—often aluminum foil or tantalum.

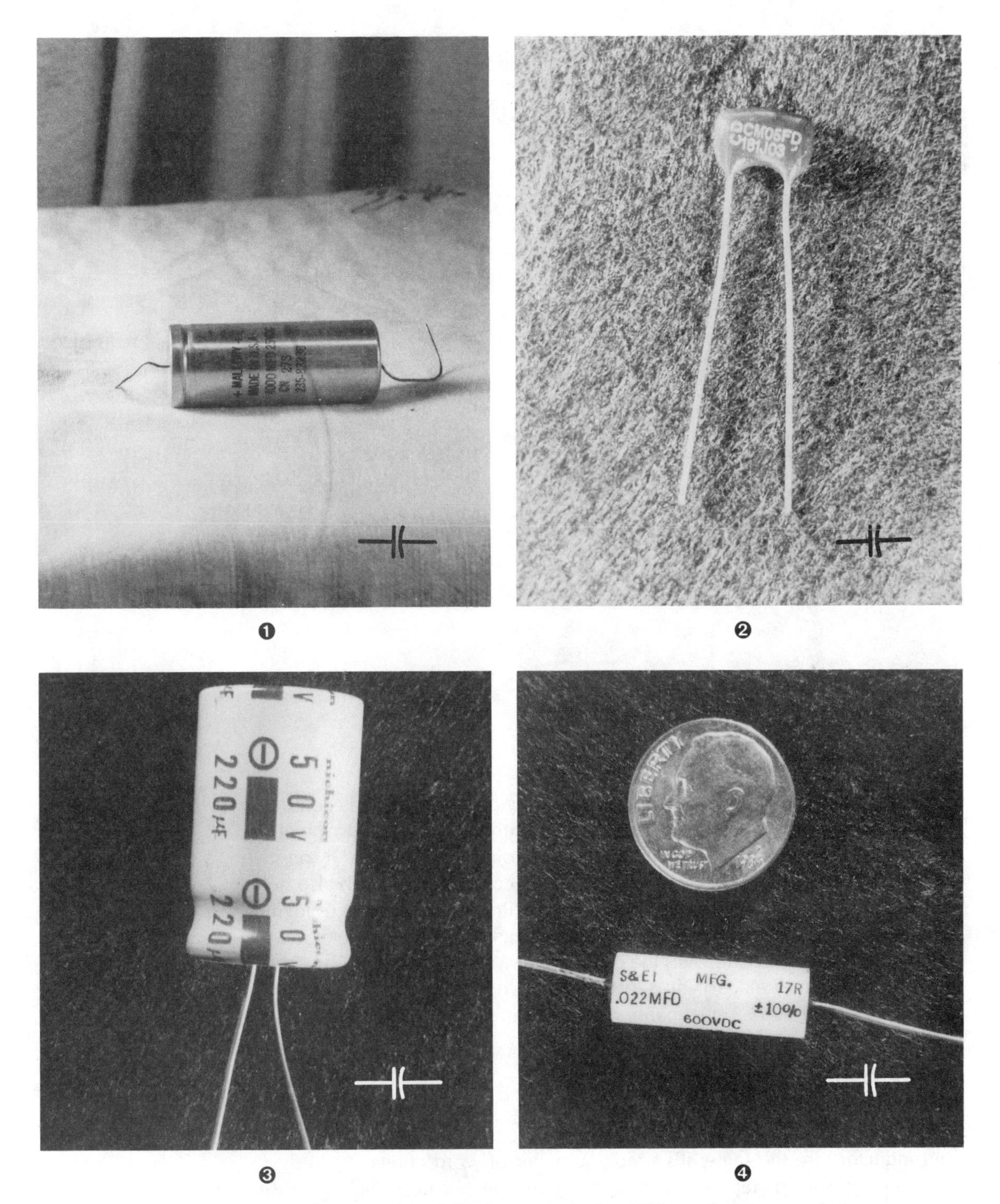

Fig. 1-8. Five different styles of fixed capacitors, but all have the same symbol.

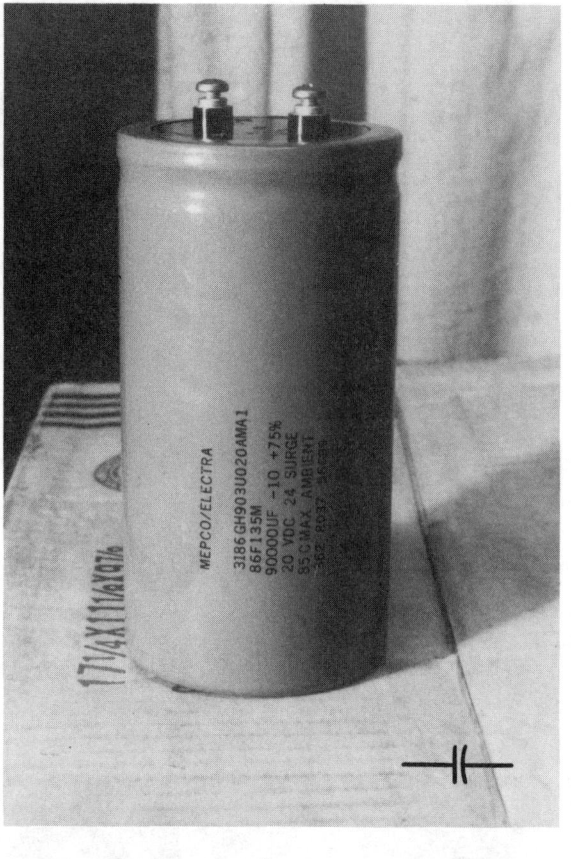

Fig. 1-8. Continued.

❺

An electrolytic is designed mainly for filtering purposes and it is *polarized*, meaning that one terminal of the capacitor is positive with respect to the other, like a magnet. The polarized terminals on an electrolytic are identified by positive (+) and negative (−) signs stamped right on the component, which also helps to identify the fact that it is an electrolytic capacitor. Most schematics identify an electrolytic by placing the plus and minus signs on the schematic next to the symbol, which also indicates the circuit polarity connections for the capacitor. Another type of electrolytic capacitor is the *nonpolarized* unit, however, these are marked in the same manner and used in special circuit applications.

COILS AND CHOKES

If you were to take a pencil and wrap several turns of wire around it, you would have a coil. It could also be called an *inductor* because, basically, coils and inductors are the same; it's simply a matter of word choice.

Some people prefer the word inductor to coil because the property or electrical characteristic of a coil is called *inductance*, just as the action of a resistor is called resistance. An inductor (coil, choke) opposes any change in current by inducing an opposing voltage. Coils (or inductors) and chokes are the same in this respect. Coils are designated by the letter L and the term used to indicate the amount of inductance is the *henry*. (See Fig. 1-9).

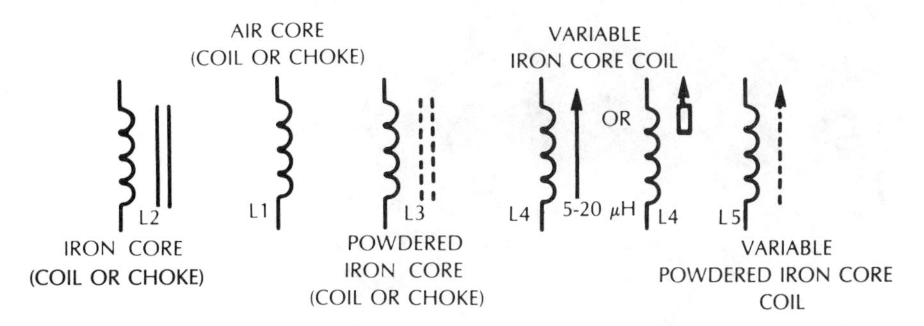

Fig. 1-9. Schematic symbols used to identify coils.

A coil or choke in radio circuits usually has a value less than one henry, so the terms *milli* or *micro* (abbreviated mh or μH) are used. For example, next to a coil or choke on the schematic you will likely have the reference number, say it's L12, and the value: 10 mh. The L12 means it is the 12th inductor and 10 mh means that it has a value of 10 thousandths of a henry. The micro symbol (μ) means the same as in the case of the capacitors—one millionth. In this case, a 10 μh coil would be read as 10 microhenrys. Figure 1-10 shows several coils and Fig. 1-11 shows a commonly used choke.

Fig. 1-10. (A) Large coil with schematic symbol.

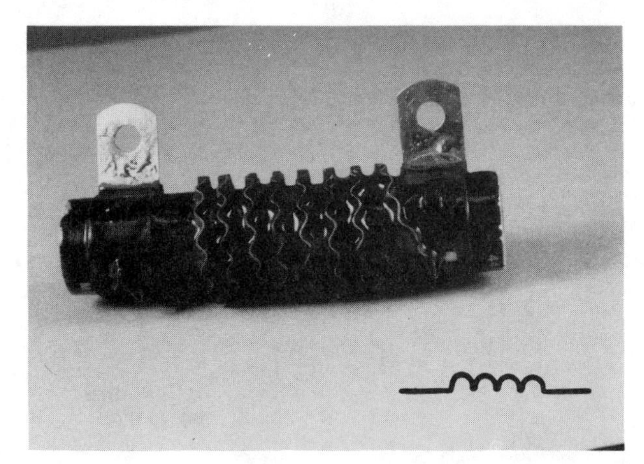

Fig. 1-10. (B) Medium-sized coil with schematic symbol.

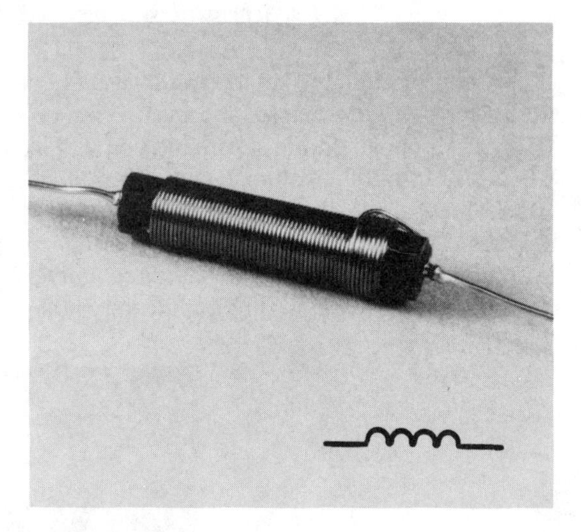

Fig. 1-11. A small rf choke. Note symbol is the same as for a regular inductor.

In actual circuit applications, the amount of inductance of a coil or choke depends on many things, such as the number of turns of wire, the size of the wire, the size of the form, and the type of core. As suggested earlier, if you wound a coil of wire on a pencil, you would have an inductor with a certain amount of inductance. If you were to wind the same coil on an iron rod, the inductance would be much greater. Also, if you wound the coil on a form and placed an iron slug inside the form which could be moved into or out of the coil, you would have a coil with a variable inductance. However, chokes are not variable.

TRANSFORMERS

Transformers differ from coils in that they are a combination of two or more coils positioned close to each other, physically, which provides the means to transfer energy from one coil to another. As shown in Fig. 1-12, the two coils of a transformer are labeled *primary* and *secondary*. Just as in the case of coils or inductors, a transformer can have an air core, a fixed iron core, or an adjustable iron core. When an ac signal is connected across the primary coil, the resulting magnetic field causes or *induces* a voltage in the secondary winding and so the signal is transferred.

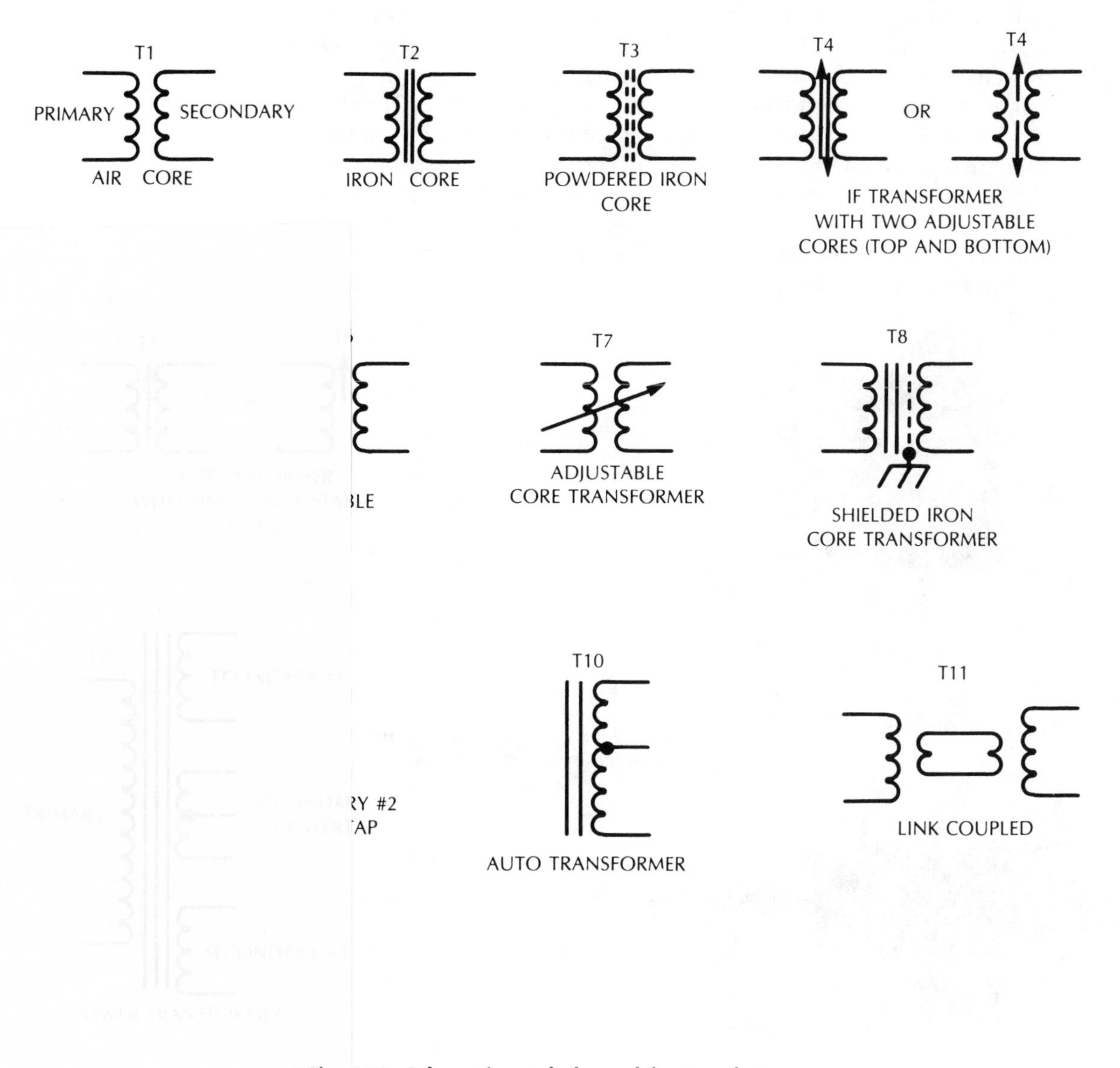

Fig. 1-12. Schematic symbols used for transformers.

Many things actually take place during this transfer, but as a matter of information, the signal appearing at the secondary will not be exactly like the signal coming into the primary because of losses in the transformer windings, the difference in the number of turns between the primary and secondary, etc.

Transformers are labeled with the letter T on a schematic diagram. But unlike capacitors, resistors, and coils, their values are not normally given (except perhaps in the parts list). Because transformers have inductance just as coils, the primary and secondary coil windings will have some value in henrys, however, these values are seldom included on a schematic.

There are many types of transformers for many applications. Transformers are used in audio, i-f (intermediate frequency), rf (radio frequency), and power-supply circuits. Transformers designed for audio and power-supply circuits generally have color-coded leads indicating primary and secondary windings. Like the resistor and capacitor color codes, transformer coding is standardized by the EIA and appear in Appendix B. Two common types of transformers are shown in Fig. 1-13.

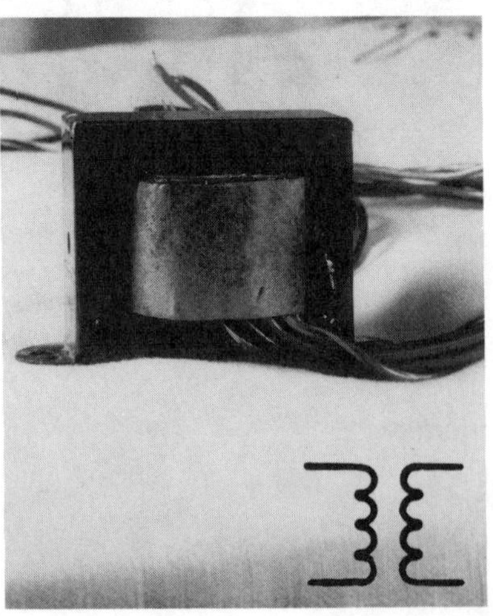

Fig. 1-13. These small transformers are used extensively in electronics.

BATTERIES

You use many kinds of batteries in your everyday activities. Your automobile, transistor radio, and flashlight all operate from batteries. It is a source of power in the form of a dc voltage that comes from cells that change chemical energy into electrical energy.

Batteries are classified as being either *primary* or *secondary* and they can be either *wet* or *dry*. The batteries in your flashlight are examples of dry primary cells, as are the batteries in your transistor radio. Primary batteries are made to develop a certain amount of energy, and when that energy is used up, you install new batteries. Your automobile battery is a secondary battery—a lead-acid wet cell—and it can be recharged with energy so it will last for years.

All batteries are represented by the same schematic symbol. Fig. 1-14 shows the symbols for a single-cell battery and a multi-cell battery, the latter being nothing more than a group of single cells connected in series. The symbol does not show the type of battery, but for practical purposes it is not important. The voltage and polarity shown are important, however.

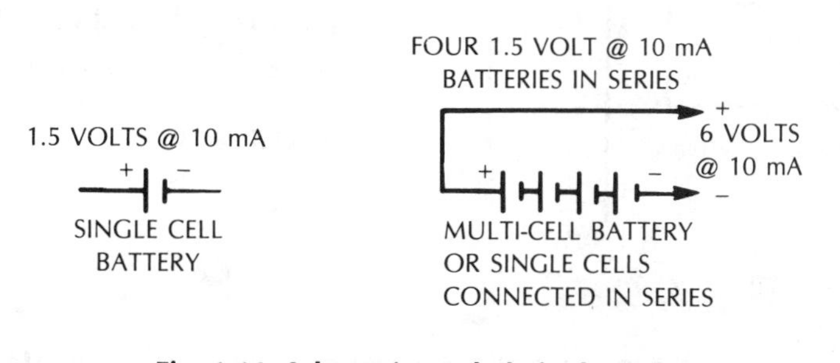

Fig. 1-14. Schematic symbols for batteries.

If your radio schematic or instructions call for a 9-volt battery, you wouldn't install one that is 22 volts, and you surely wouldn't get much music out of the radio if you used a 1.5-volt flashlight battery instead of a 9-volt battery. No matter what you use a battery for, it must also be installed with the right polarity or it will not work.

Batteries have two poles or terminals—one positive (+) and one negative (−). As illustrated in Fig. 1-14, the longer vertical line is the positive terminal while the short line is negative. Even though the battery symbol itself indicates the polarity, the plus and minus signs are often included on the schematic. When two or more batteries are connected in series, their respective voltages add, as shown by the schematic symbol for a multi-cell battery.

An automobile battery is an example of several cells connected in series to provide 12 volts. Flashlights using two or more single cells stacked in the holder are also series-connected.

In most applications, a specific battery voltage is required, but the current needed might be more than one cell can supply without putting an excessive drain on it. In this application, batteries can be connected in parallel as shown in Fig. 1-15(A). This arrangement provides the same voltage as one battery, but now their respective current capabilities are added. When using this type of connection, the batteries should all be the same voltage rating to prevent excessive current drain on the battery having a lower voltage. Batteries can also be connected in series and parallel combinations to provide both more voltage and current as shown in Fig. 1-15(B).

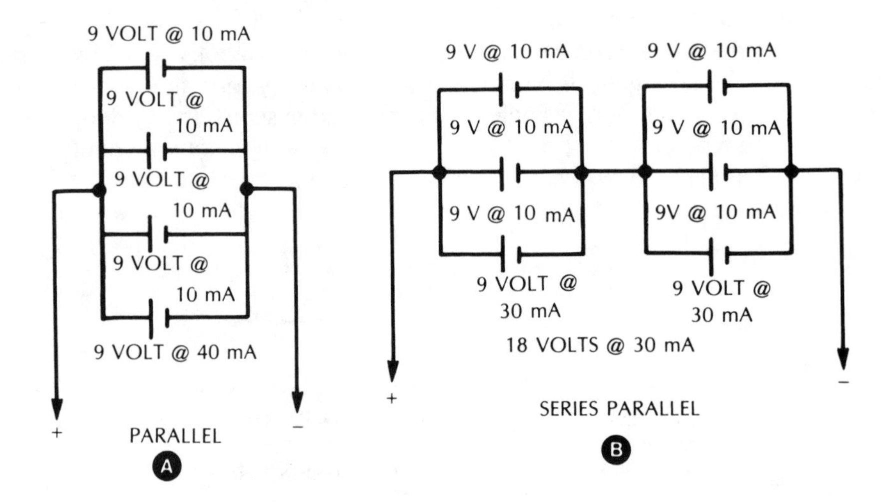

Fig. 1-15. Connecting batteries in parallel (A) increases the available current. More voltage and current can be obtained by a combination series-parallel hookup (B).

As mentioned previously, the schematic symbol does not tell you what kind of battery it is. But certain batteries are used in given applications primarily because of their ratings, size, and cost. Figure 1-16 shows several types. For example, you could use an automobile battery to operate a flashlight or a transistor radio, but you would have a tough time carrying it around. Moreover, a flashlight or transistor radio doesn't need the current available from an automobile battery—a small dry cell works just as well and lasts long enough to make the cost of replacing these cells periodically a relatively minor expense. Dry cells, because of their chemical makeup, cannot be recharged.

A type of rechargeable battery which can be used in flashlights, portable radios, and electronic equipment is the nickel-cadmium battery. Mercury and alkaline cells are other dry-cell batteries now available for low-power operation, such as in flashlights and transistor radios. The mercury battery is more expensive than the ordinary dry-cell flashlight battery, but it can be stored for a much longer time and its output voltage is more constant. The alkaline battery is also more expensive than the ordinary carbon-zinc dry cell, but it is capable of a much longer life.

Fig. 1-16. Nickel-cadmium (nicad) batteries can be recharged hundreds of times. (A) "AA" cell, (B) "D" cell, (C) 9-volt battery (courtesy of Radio Shack, A Tandy Corporation).

SWITCHES

Every time you turn on a radio or start your car, you are making an electrical connection through a switch. The courtesy light in your automobile comes on because a door switch closes, as does the light in your refrigerator. Pushbutton switches are used in such applications. Because the switch completes only a single circuit between the light and a power source, it is known as a single-pole switch. The schematic symbol is shown in Fig. 1-17. Switches are available in toggle, knife, pushbutton, slide, and rotary selector or wafer types.

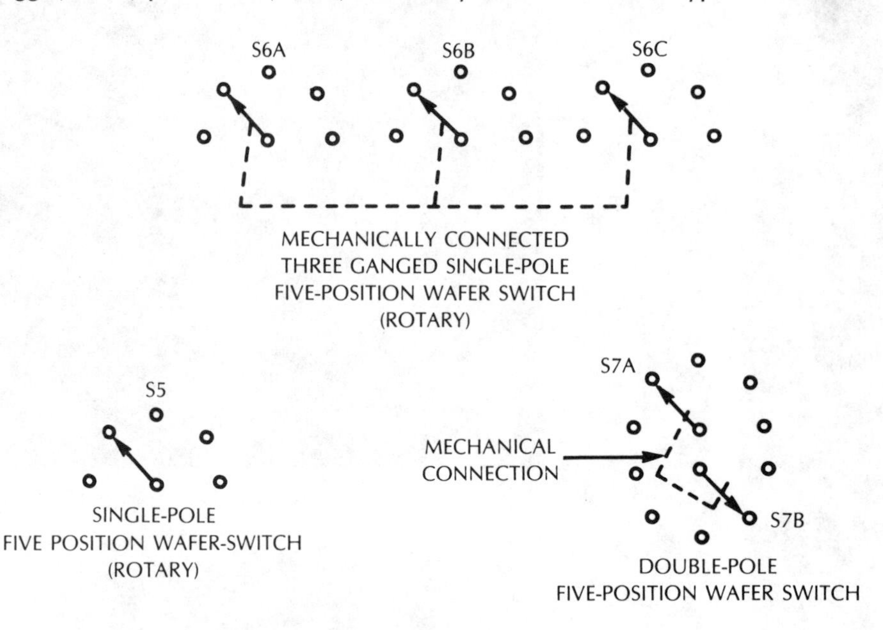

Fig. 1-17. Schematic symbols used for switches.

Many switches have multiple contacts to complete many circuits at once or to connect one point in the circuit to any one of several circuits. Regardless of the type of switch, they are all identified on a schematic with the letter S, for example: S1, S2, or S3. Ganged switch sections are labeled S1A, S1B, etc. The schematic symbol for multiple-contact switches is also shown in Fig. 1-17.

As illustrated, a single-pole single-throw switch is the most basic and will be the type you are likely to find as a part of a radio or TV volume control. A single-pole double-throw (SPDT) switch is used to provide a connection from one main contact to either of two circuit contacts. A double-pole double-throw (DPDT) switch is actually two single-pole double-throw switches mechanically connected or ganged. DPDT switches (as well as others) are available as slide-operated, toggle, knife, etc., depending on the application. Figure 1-18 pictures slide and toggle type switches. In electronic circuits that require multiple connections, a rotary selector or wafer switch is normally used (Fig. 1-19). The combinations and applications for this type of switch are unlimited. Wafer switches can also be ganged or mechanically connected.

Fig. 1-18. Slide and toggle switches with symbol.

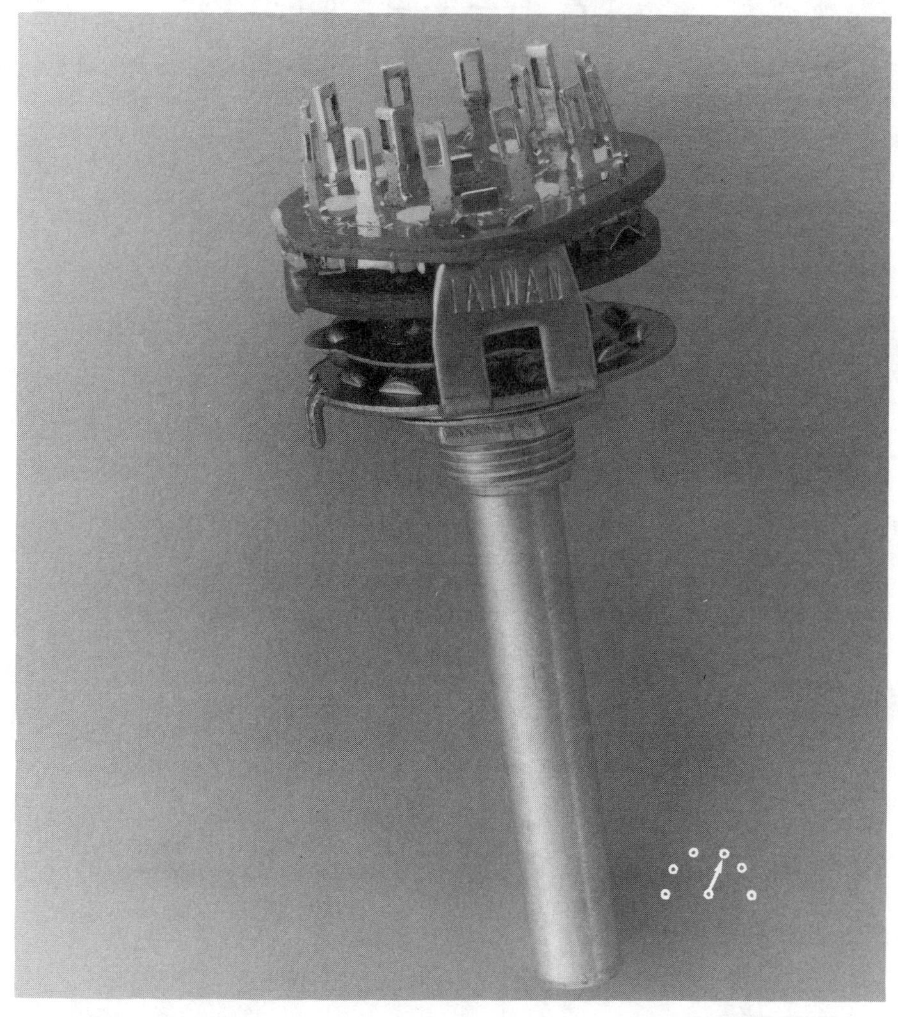

Fig. 1-19. A multiposition rotary switch (courtesy of Radio Shack, A Tandy Corporation) and schematic symbol.

RELAYS

In the discussion of switches, we said that a switch is used to complete a circuit or circuits. A relay is a switch too, but it is operated electronically or electrically—not manually. We can best illustrate the way a relay works by referring to the diagram and schematic symbols shown in Figs. 1-20 and 1-21. As shown, the relay has an iron core coil and an *armature* that is mechanically connected to a set of contacts. The spring attached to the armature pulls the contacts open when the relay current is interrupted. The schematic symbol is basically the same for all relays. The only variation is in the number and arrangement of the contacts.

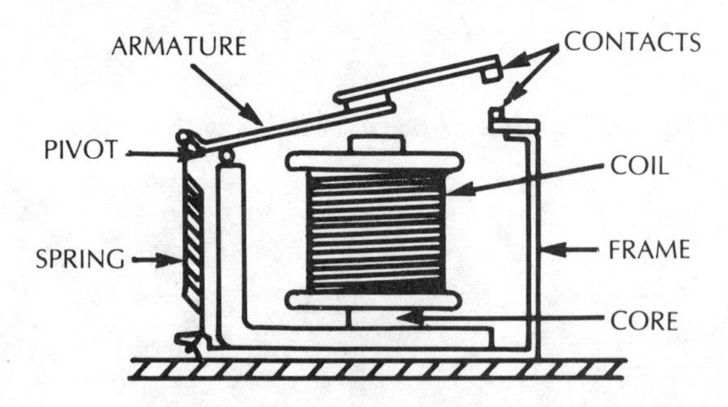

Fig. 1-20. Drawing of a typical relay, showing principal parts.

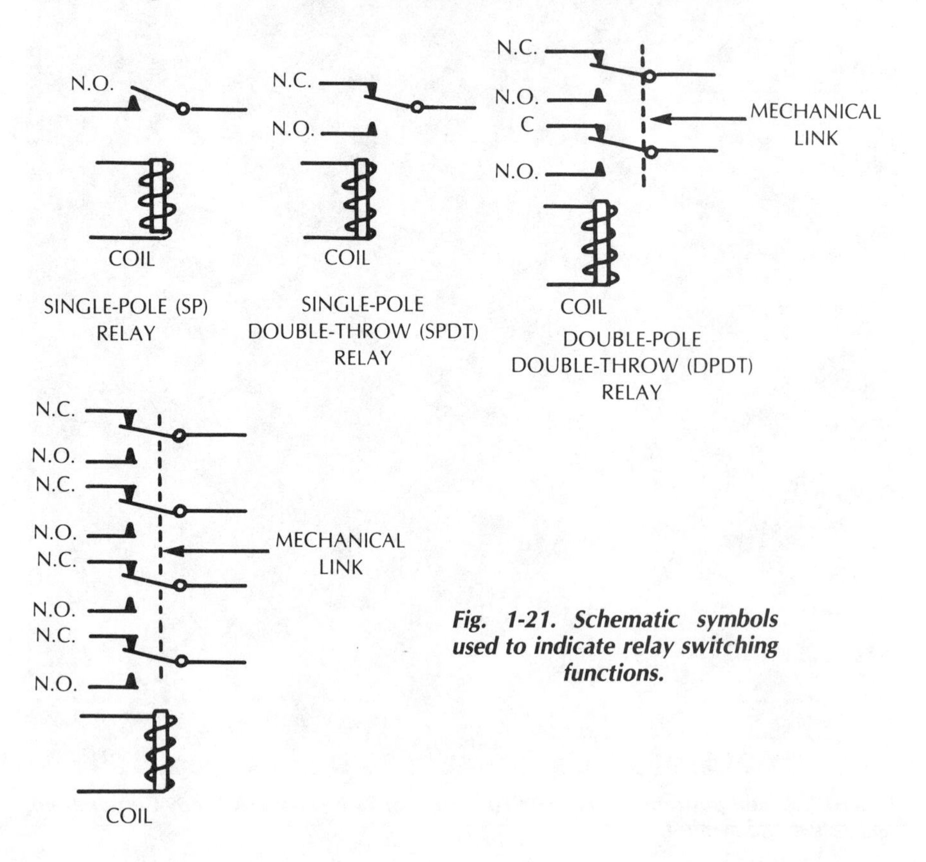

Fig. 1-21. Schematic symbols used to indicate relay switching functions.

The reference letter designation for a relay is K; the contacts themselves are usually numbered, too. The single-pole relay (SP) has only one set of contacts. Relay contacts can be either made to open or close, depending on how they are arranged. Contacts that open when the relay is not operating or energized are called *normally open* and labeled NO on the schematic symbol. Contacts that are closed when the relay is not energized are called *normally closed* and labeled NC on the schematic.

As illustrated, a relay can have many contacts to perform several switching functions at one time. The contacts are mechanically linked so they all close (or open) at the same time. Relays are available for hundreds of applications. Two of the more common types are shown in Fig. 1-22. A solenoid is a relay, but instead of a contact, the magnetic field moves a plunger or arm that is used to perform some mechanical function.

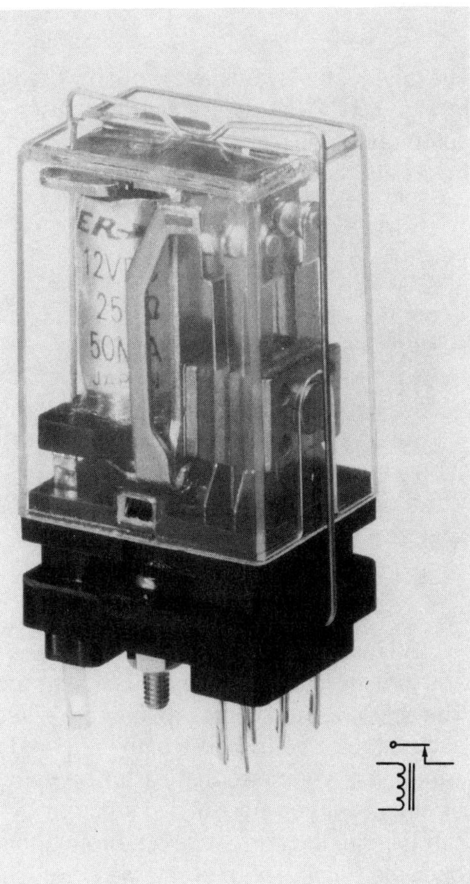

Fig. 1-22. You can see the contacts in this type of relay, however, they are often found in opaque cases (courtesy of Radio Shack, A Tandy Corporation).

FUSES

A fuse is a relatively simple device, but it performs many important functions in your home, automobile, TV, and other appliances. The schematic symbols for a fuse are shown in Fig. 1-23.

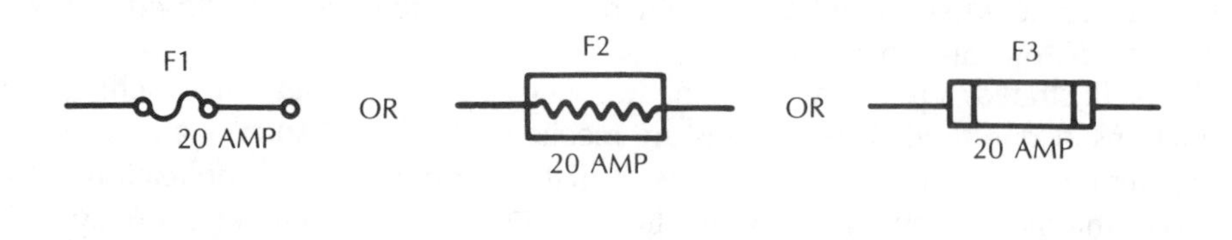

Fig. 1-23. Fuses are identified by these schematic symbols.

The types of fuses you have probably used and are most familiar with are shown in Fig. 1-24. These glass cartridge fuses are normally used in TV, hi-fi, and automobile fuse panels, while the screw-in type is generally found in home fuse boxes and heavy appliances.

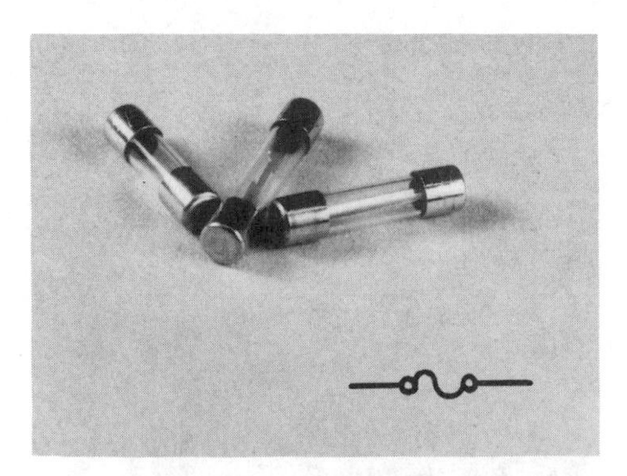

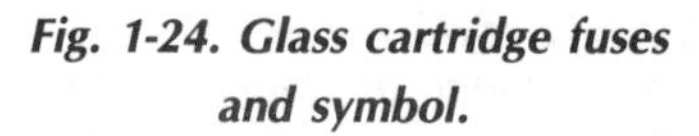

Fig. 1-24. Glass cartridge fuses and symbol.

All fuses operate (or open) on the same principle—heat. You know that when you turn the lights on in your home, they are drawing a certain amount of current. The ac outlets and ceiling fixtures are wired in a number of branch circuits, and each is protected by a fuse of the required rating. If you were to look at the fuse panel, you would see that each fuse perhaps takes care of two or three rooms or circuits. If you overloaded a circuit by plugging in more appliances than it can handle, the current going through the fuse will melt the fuse element and open the circuit. This prevents the excess current from damaging any components in the circuit. The fuse element is nothing more than metal alloy with a low melting temperature. The higher the fuse rating, the larger the element.

The fuses in your television set and automobile work the same way, except they are physically different. Because fuses are designed to protect an appliance or wiring, it is dangerous to replace a fuse with one rated higher than the original.

The fuses used in home fuse panels are screw-in types. Cartridge-type fuses shown in Fig. 1-25 are plug-in or clip-in types made to fit into specially designed holders or clips (Fig. 1-25). Some fuses are wired into the circuit and have leads connected to them for this purpose. The trend toward miniaturization and the use of transistors has created a need for miniature versions of pigtail fuses.

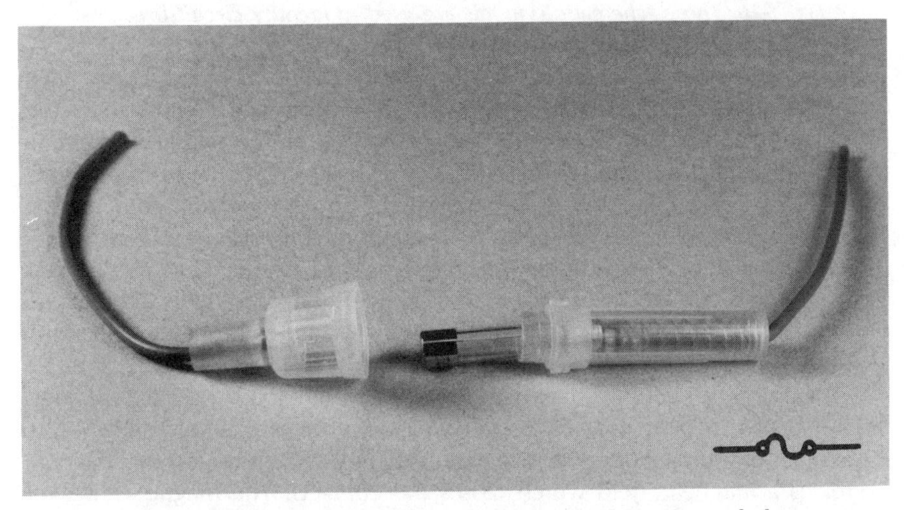

Fig. 1-25. A spring-loaded fuse holder and schematic symbol.

Another commonly used fuse is the *slow-blow* fuse. It has a special element with a built-in time delay for use in circuits where there may be a momentary high surge of current when the equipment is first turned on. Motor-starting circuits and some electronic equipment power supplies have an initial surge or overloaded current that is higher than the normal operating current. Because the surge is sometimes two or three times the normal current, it would be dangerous to fuse it at that rating. So, we use a slow-blow fuse that will hold until the surge passes and then safely protect the circuit at its normal current rating.

CIRCUIT BREAKERS

The schematic symbols for circuit breakers are shown in Fig. 1-26. Circuit breakers have become popular replacements for fuses in many applications such as in home wiring, television sets, and industrial circuits. Circuit breakers may be either thermally or magnetically operated.

A circuit breaker differs from a fuse in that it does not burn out or need to be replaced after an overload condition. It is an electromechanical device.

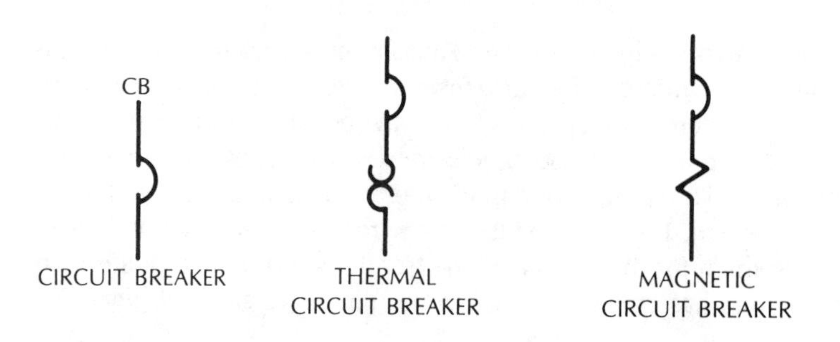

Fig. 1-26. These schematic symbols are used to identify circuit breakers.

Thermal circuit breakers are made with a special type of metal strip, called *bimetal*, that touches a contact arm and completes the circuit. This bimetal strip has a specific rating and as long as the current is within limits it will remain in position against the contact arm. If the current goes over the normal rating, heat will cause the bimetal strip to bend away from the contact arm and open the circuit. Depending on what kind of thermal circuit breaker it is, it will be reset mechanically or it will reset itself automatically. The mechanical reset units are pushbutton operated. Breakers that reset themselves automatically have no button. As the bimetal strip cools, it automatically returns to the closed-circuit position.

The magnetic circuit breaker does the same kind of job a thermal unit does, only it operates somewhat like a relay. Its symbol is shown in Fig. 1-26. Remember our explanation of a relay coil. When current flows through the coil, it creates a magnetic field which closes the contacts. The magnetic breaker operates about the same, except instead of contacts, it uses a plunger, like a solenoid. The plunger makes contact to complete the circuit and is held in place by a spring that has just the right amount of tension to overcome the pull of the relay when normal current is flowing through the relay coil. When an overload condition comes along and more current flows through the coil, it exerts a stronger magnetic pull. When the pull gets strong enough, it overcomes the spring tension and pulls the plunger away from the contact which opens the circuit. The circuit breaker has to be reset manually. This type of breaker is usually found in homes and industry.

2
Transducers, Indicating Components, and Miscellaneous Components

In addition to the discrete, (individual) circuit components covered in Chapter 1, there are those that change sound or mechanical movement into electrical facsimiles (*transducers*) and vice versa, and those that provide aural or visual indications, such as meters. Although some of these components are normally found only in special equipment, a complete knowledge of what each item looks like, both physically and on a schematic, will help you better understand a schematic diagram.

CRYSTALS

The exact definition of a crystal can be difficult to understand without some background in physics. Briefly and simply, a crystal is made from a solid, such as quartz, barium titanate, or Rochelle salt. These particular solids are used because they have *piezoelectric* properties. A crystal is cut from quartz somewhat like cutting a precious gem. When a crystal is cut in a certain way, it will vibrate at a specific frequency when voltage is applied in an electronic circuit. Therefore, crystals can be used to control the frequency of many types of oscillators, such as those found in radio transmitters, signal generators, color TV sets, FM receivers, and hundreds of other circuits. The reason crystals are so popular for frequency control is because their vibrating frequency is very stable, especially under controlled voltage and temperature conditions.

Crystals are shown on schematic diagrams by the symbol illustrated in Fig. 2-1. The schematic letter designation is the letter X, although the letter Y is sometimes used.

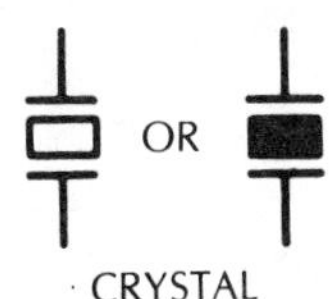

Fig. 2-1. Symbols used to depict a crystal.

Figure 2-2 shows the physical appearance of several types of crystals. The most commonly used crystals in consumer electronics equipment are of the plug-in or wired-in miniature type. Some communications equipment uses a larger type of plug-in crystal while other equipment uses heated crystal units. The heated or oven-controlled crystal unit is generally built in a sealed container that plugs into a socket supplying heater voltage and crystal connections. In tube-type equipment, the crystal heater, or *oven*, is similar to a tube filament and, it can receive its voltage from the filament line.

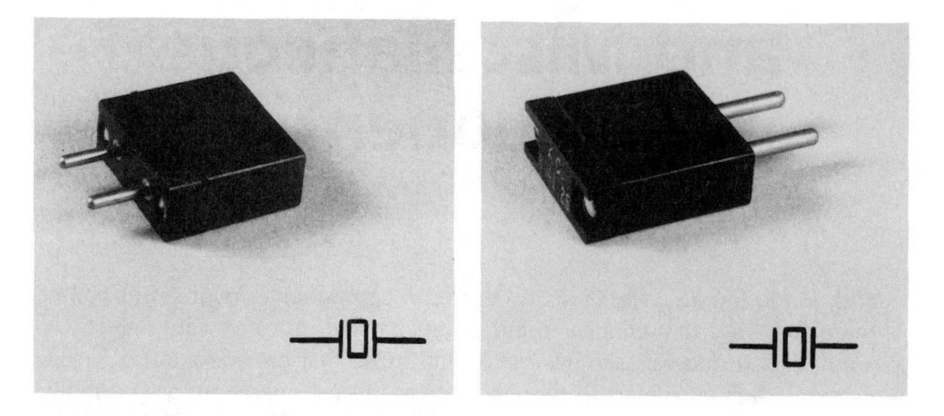

Fig. 2-2. Crystals come in different case styles like these.

Crystal ovens often have a built-in thermostat consisting of two bimetal strips as in the circuit breaker discussed in Chapter 1. These strips are contacts which, when closed, operate a heater-voltage control circuit. When the operating temperature is exceeded, the strips or contacts separate and open the heater supply. Then the bimetal contacts cool until they again close and complete the heater circuit.

FILTERS

The term *filter* should be familiar. Your automobile engine has an oil filter, and a washing machine has a lint filter. They all do the same thing—they allow only certain things to pass. Electronic filters allow certain frequencies to pass. Filters come in many styles and designs for as many applications. The scope of this book doesn't cover their respective characteristics, but the schematic symbols for most of the various types should be helpful.

Generally, filters are made from a network of resistors, inductors, capacitors, or a combination of these, depending on the application. Crystals are also used for filter networks, as are mechanical devices. Filter networks using resistors, inductors, or capacitors are termed as *low-pass*, *high-pass*, or *band-pass* filters. These are shown in Figs. 2-3 and 2-4.

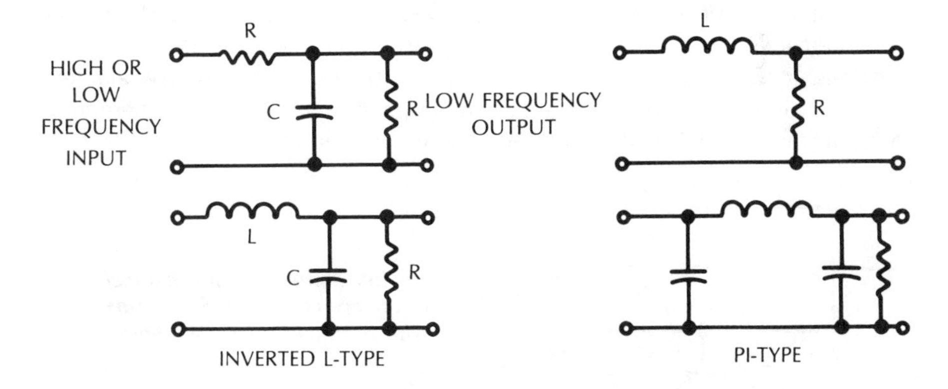

Fig. 2-3. Typical low-pass filter networks.

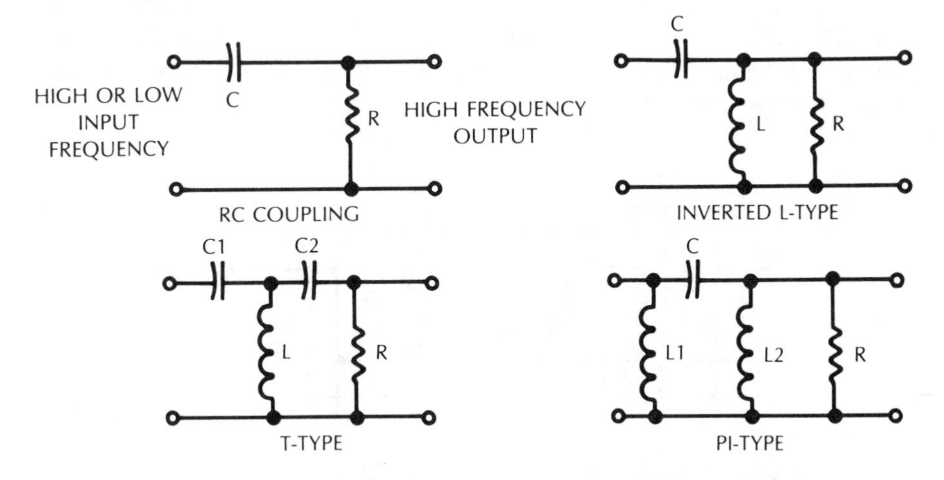

Fig. 2-4. Basic high-pass filter networks.

The entire network making up the filter circuit is sometimes enclosed in a shielded case (as shown on the schematic diagram in Fig. 2-5) by dashed lines around the circuit.

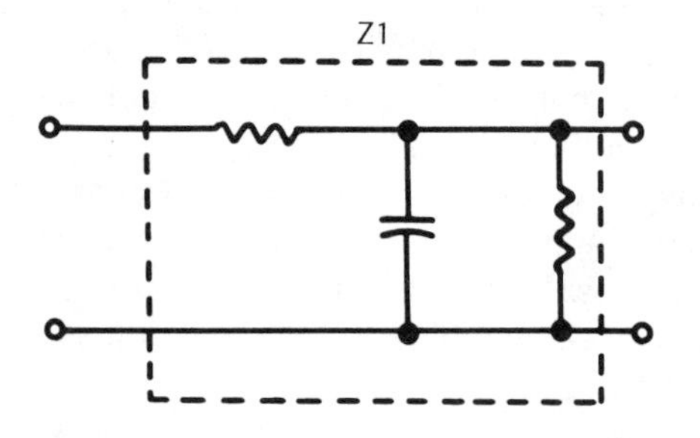

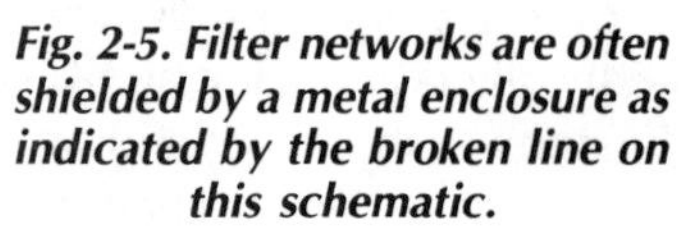

Fig. 2-5. Filter networks are often shielded by a metal enclosure as indicated by the broken line on this schematic.

A mechanical filter is almost always designed as a sealed unit and shown on the schematic simply as a box with input and output terminals (see Fig. 2-6). It is designated as FL and its frequency is given as shown. Generally, crystal filters are also sealed units, as their tolerance and design make them crucial. Attempting to repair them in the field is normally not recommended.

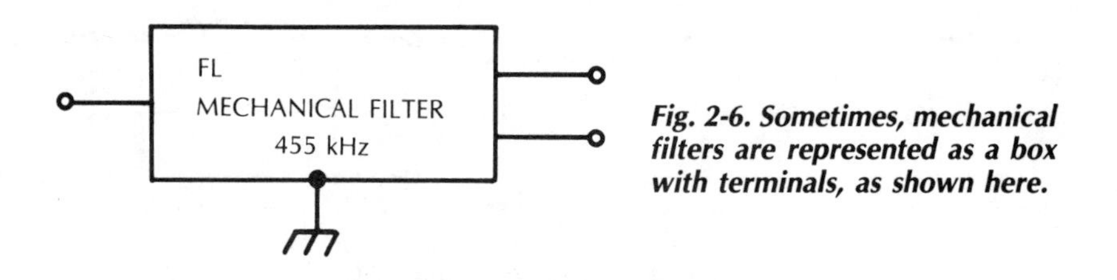

Fig. 2-6. Sometimes, mechanical filters are represented as a box with terminals, as shown here.

The schematic symbol for a crystal filter may be a box identified in the same manner as a mechanical filter, or it may show actual circuit connections (see Fig. 2-7). Crystal and mechanical filters are found in communications transceivers where an extremely high degree of selectivity is required, especially in single-shielded transmitting and receiving units.

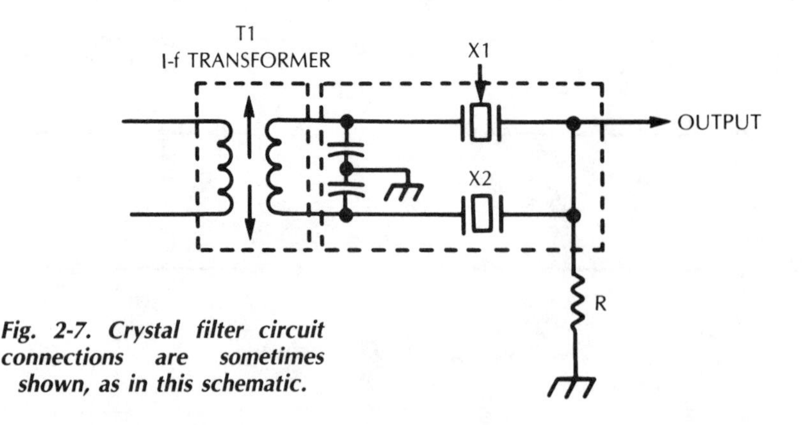

Fig. 2-7. Crystal filter circuit connections are sometimes shown, as in this schematic.

HEADPHONES

Headphones (sometimes called earphones or headsets) are normally not shown on a schematic diagram of a particular piece of electronic equipment. But headphone accessory jacks are built into many of the products we use every day, such as our TV, portable radio, stereo, and many communications transceivers. The schematic symbols for the various types of headphones and a set of stereo headphones are shown in Fig. 2-8.

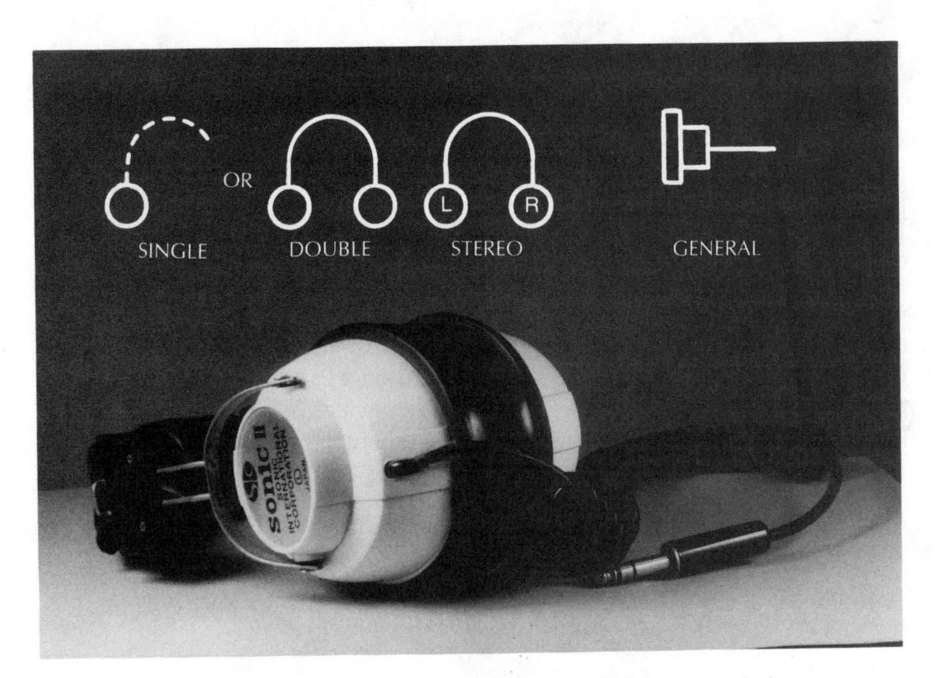

Fig. 2-8. Stereo headphones and various headphone symbols.

A headphone may be classified as crystal, dynamic, or magnetic, depending upon the construction of the *transducer,* or sound-reproducing element. The schematic symbol does not indicate this, but a footnote may be included if a specific type of headphone is required. Generally, a dynamic headphone will be used where high fidelity and low impedance (4 to 8 ohms) is necessary. Magnetic headphones are high impedance, usually 2000 ohms or more, and can be placed directly in the plate circuit of an amplifier tube or the collector circuit of a transistor. Most headphones are available in either single or double units.

LAMPS

Lamps are as numerous as snowflakes in a November storm, but the schematic symbols, fortunately, are few. The most common types are the incandescent, neon, and the readout lamps. The incandescent lamp is the type used in home lighting, flashlights, and to illuminate the dial of your radio or TV set. It is shown schematically in Fig. 2-9.

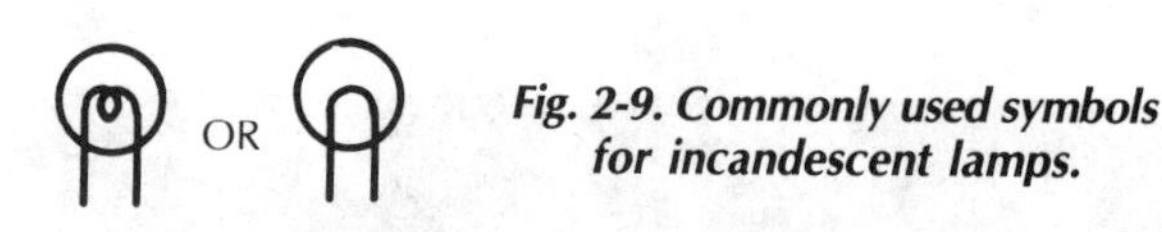

Fig. 2-9. Commonly used symbols for incandescent lamps.

Incandescent lamps come with many types of bases and in many voltage ratings. The base can be a screw-type, bayonet, candelabra, or have wire leads. The letter designation for a lamp is the letter L.

Neon lamps have more specific uses because they are gas-filled and come on only when a certain voltage is applied. When a neon lamp turns on, it is said to be *ionized.* The voltage necessary to ionize a neon lamp is approximately 65 volts or more, and they are most often used as indicators in specialized equipment. However, you will also find these devices used in relaxation oscillators and waveform-generating circuits. The schematic symbol for a neon lamp is shown in Fig. 2-10.

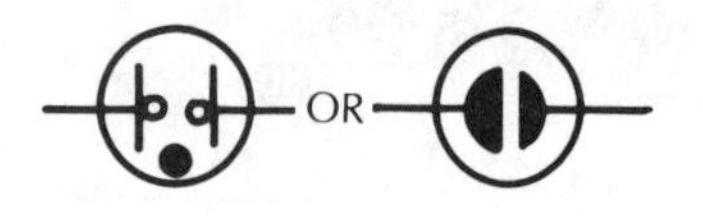

Fig. 2-10. Both of these symbols are used for a neon lamp.

A readout lamp or indicator is another special type of device. It is used as a numerical display indicator in digital counters, voltmeters, and other similar units. The schematic diagram for a readout indicator is shown in Fig. 2-11. It is much like a tube in appearance, but has many cathodes. The cathodes are made in the physical shape of a number that lights or glows when voltage is applied to it. Many of these readout units are gas-filled neon, while others use incandescent lamps in a special arrangement to indicate numerals or letters.

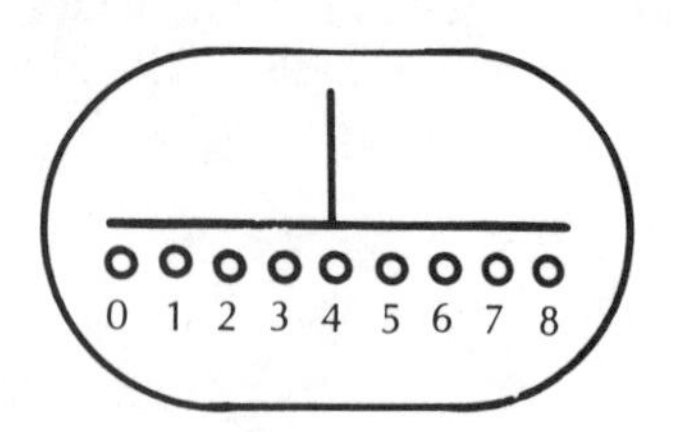

Fig. 2-11. Readout lamps are designated by this symbol.

METERS

Most schematics simply show a meter as a circled letter that designates whether the meter is for measuring voltage (V), current (A), signal level (dB), etc. The polarity at which the meter is connected in the circuit is given on the respective terminals.

In most circuits, meters are installed to monitor specific values of voltage, current, or signal. In some circuits, such as in transmitters, an ammeter may be used to monitor several circuits by connecting the meter to these circuits through a switching arrangement. In tape recorders, however, a signal-level meter is normally connected to monitor the incoming level on only one channel. Stereo recorders have one level meter for each of the two channels. Some stereo FM receivers also use a signal level meter to indicate when the station is correctly tuned. This type of meter shows a balanced condition when the receiver is centered on the station.

One of the most common uses for meters is in test instrument circuits such as tube testers, volt/ohmmeters, vacuum-tube voltmeters, signal strength meters, and many more (Fig. 2-12). The meters in such instruments are often used for several functions through a variety of switching arrangements. The basic meter is *shunted* (bypassed) or in series with resistors which allow the meter to read higher values of current or voltage than the meter would normally be capable of without shunt or *multiplier* resistors.

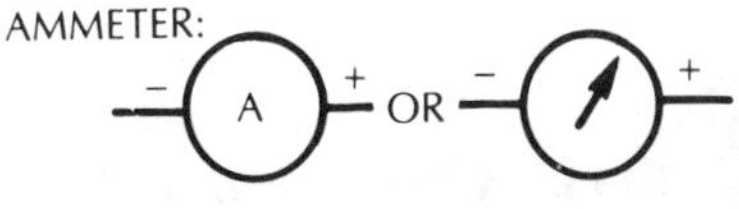

AMMETER:

– A + OR – +

A = AMMETER
mA = MILLIAMMETER
V = VOLTMETER
dB = DECIBEL METER

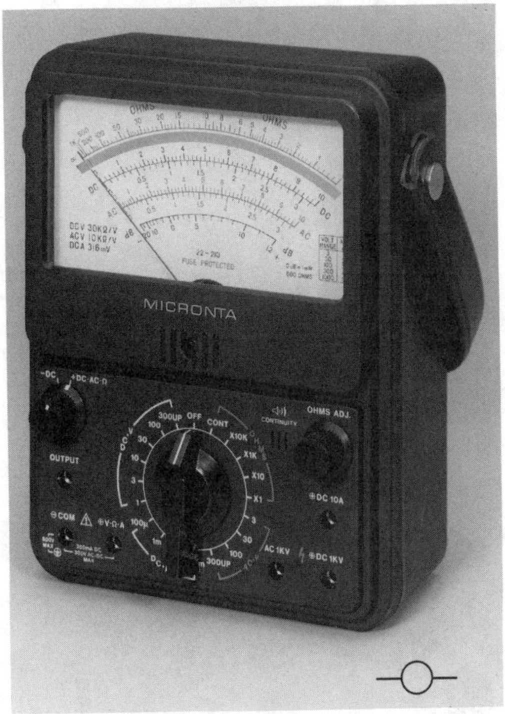

Fig. 2-12. This Micronta multimeter can measure voltage, current, and resistance (courtesy of Radio Shack, A Division of Tandy Corporation). One of the above abbreviations goes in the circle to designate which type of meter is being used.

MICROPHONES

Earlier broadcasters used microphones that were quite large and some were rather awkward to look at. Today, when a modern TV interview takes place on your screen, you hardly notice the small microphones the personalities wear around their necks.

Today, microphones, in one way or another, affect everyone's life. We use them for our tape recorders, two-way radios, and public address systems in churches, schools, and business, not to mention the entertainment fields. A microphone turns sound, either music or voice, into electrical energy.

Naturally, there are many types of microphones, but again, the schematic symbols are few. As shown in Fig. 2-13, microphone symbols indicate whether the device is a general type or directional. *General* microphones may be almost any type, such as those used for radio communications, public address, and home entertainment. A *directional* microphone picks up only sounds directly in front of it. In other words, a directional microphone will not pick up noise or sound from behind. They are used by most entertainers, broadcast studios, and recording companies.

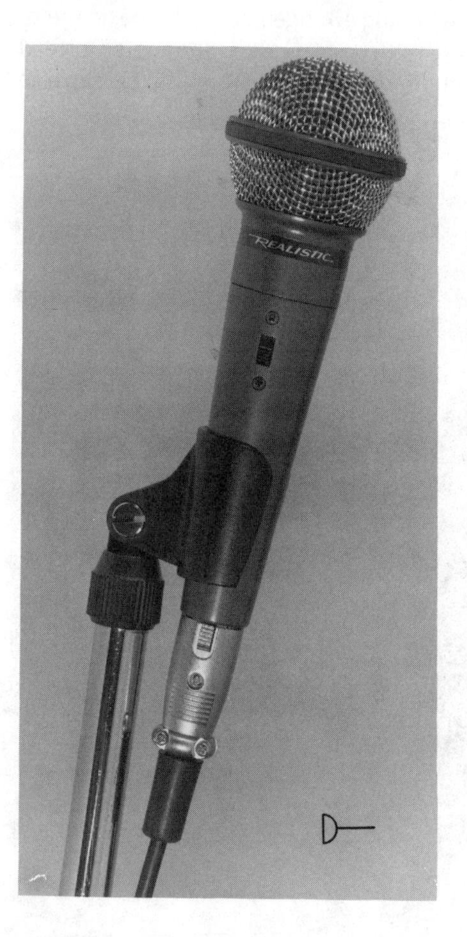

Fig. 2-13A. A unidirectional microphone for public address systems and hand-held use (courtesy of Radio Shack, A Division of Tandy Corporation) and symbol.

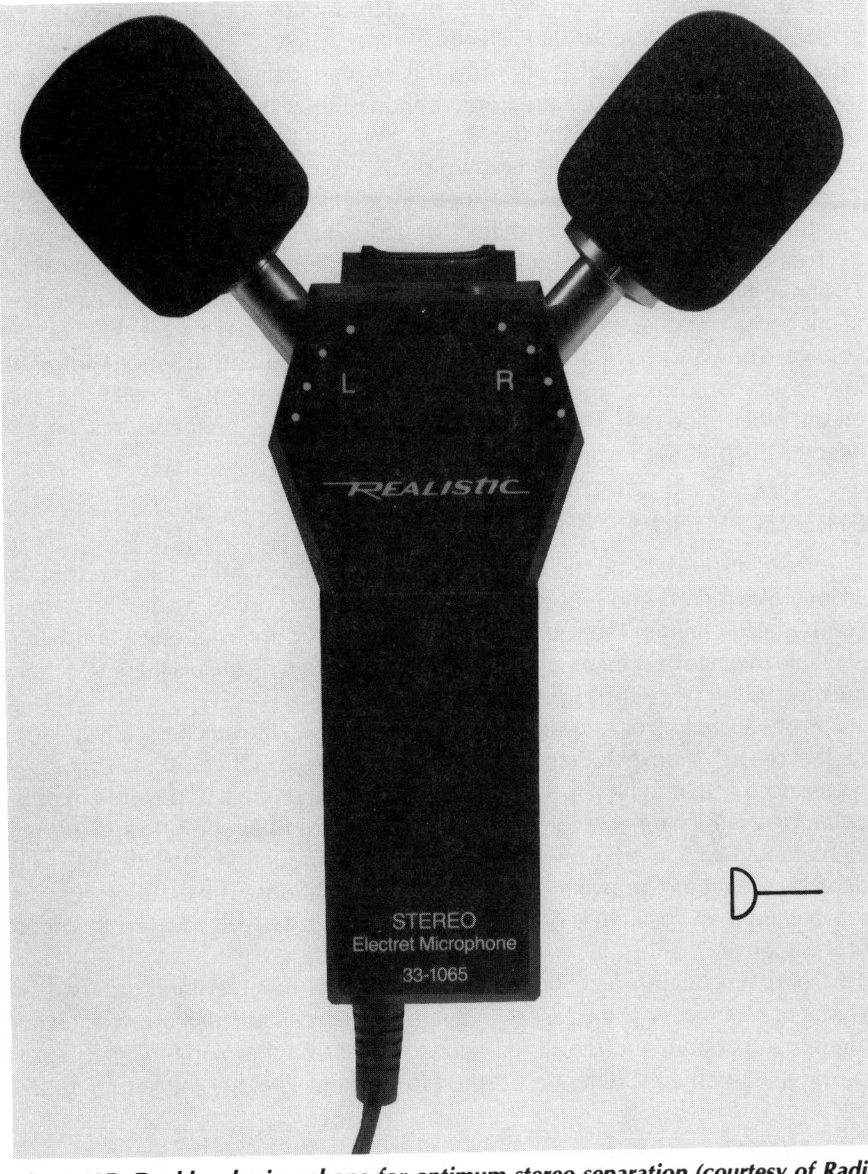

Fig. 2-13B. Dual-head microphone for optimum stereo separation (courtesy of Radio Shack, A Division of Tandy Corporation).

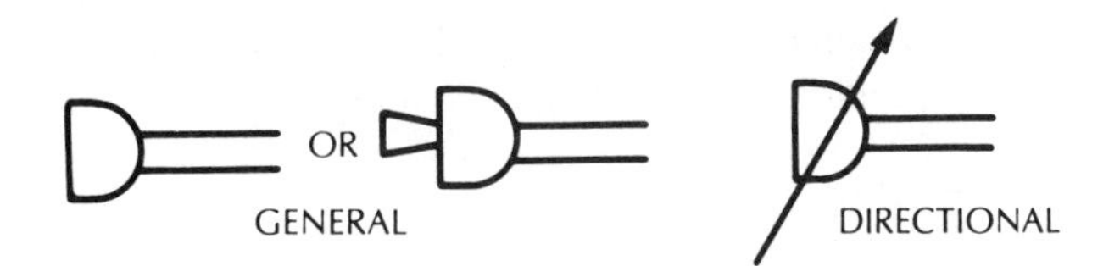

Fig. 2-13C. Microphone schematic symbols.

Even though there are many types of microphones, their job is basically the same—to change sound into electrical energy. A microphone has in its mouthpiece an element that performs this change, called a *transducer.* One of the things that makes a particular microphone better than another for a certain job is the material used in the transducer. Some of the more common types of microphones are the crystal, ceramic, dynamic, carbon, and condenser.

A crystal microphone uses a piezoelectric element, like the crystals previously mentioned, and it usually provides a fairly high output voltage. Ceramic microphones are more rugged than crystal types but operate in the same way. Dynamic types are popular in many applications, but their output voltage is relatively low. Carbon microphones are also rugged and have a very high output voltage. They also have a limited frequency response and are more common in two-way radio communications than in other applications. Condenser types, on the other hand, are used in applications where wide frequency response is needed, such as for music recording.

PHONO PICKUPS

Phono pickups or phono cartridges change the mechanical vibrations in the groove of a record into electrical signals. When a record is made, the record-cutting head changes the sound picked up by studio microphones, amplifiers, etc., into mechanical vibrations. A needle in the cutting head cuts these vibrations into record grooves corresponding to the sounds.

A phono pickup or cartridge needle simply follows the mechanical vibrations in the record groove to change them back into electrical signals that are reproduced and amplified to sound as much like the original as possible. A *monaural,* or single-channel recording, has vibrations on only one side and a single element pickup is used. Conversely, in stereo reproduction, the side walls and depth of the record groove are used to record vibrations. The dual-element stereo pickup then receives vibrations from the walls as well as from the depth of the groove.

Therefore, phono pickups are either monaural or stereo and are made of crystal or ceramic. As is the case with microphones, ceramic pickups will tolerate higher temperature and moisture. Figure 2-14 shows a moving-magnet phono cartridge. Usually, the letter designation for pickups is either the letter P or PU.

SPEAKERS

One of the most important components in any electronic device providing audio is the speaker. It provides the final result—sound. Speakers for today's electronic products are available in dozens of types and styles. Television sets, radios, hi-fi stereo, paging, and communications systems all use speakers, but the design used in each can be entirely different.

Fig. 2-14. Realistic/Shure moving-magnet phono cartridge (courtesy of Radio Shack, A Division of Tandy Corporation).

The most popular types of speakers are the permanent-magnet dynamic type, used in transistor radios, TV, and hi-fi equipment. Another type is the electrostatic. All speakers have the same basic function—to reproduce sound. Some speakers, because of their design, can reproduce only certain frequencies at certain power levels. For example, a speaker designed for a TV or radio normally does not have to deliver more than a few watts of power because that is all the circuit requires. Also, the fidelity or range of frequencies is limited by the design of the equipment. However, in hi-fi and stereo equipment, the power levels may run to almost 100 watts and the speakers must be able to reproduce sound from very low bass notes to an extremely high audio frequency range.

A speaker reproduces sound by mechanically moving the air in step with the electrical signals fed to it. Electrical signals to a permanent-magnet dynamic speaker create magnetic fields that cause the speaker cone, or diaphram, to move. Low frequencies require a large movement of air and the higher in frequency the sounds become, the smaller the air movement is.

Because fidelity and power are not crucial in low power speaker applications such as TV and radio, a standard type of dynamic speaker is often suitable. However, in high-powered, wide-frequency range stereo equipment, special speakers are used: one for low frequency, one for middle-range frequencies, and one for the highs. Such speakers are called the *woofer, mid-range*, and *tweeter*, respectively. Some systems combine speakers in a single case to provide the necessary coverage. A two-speaker combination is called a *coaxial*, which generally has a woofer and a tweeter. A three-speaker combination is called a *triaxial* and includes the woofer, mid-range, and tweeter. See Fig. 2-15.

Fig. 2-15. A three-way speaker system (courtesy of Radio Shack, A Division of Tandy Corporation) and schematic symbol.

The schematic symbols for speakers indicate only individual speakers, so a symbol for coaxial and triaxial units would have to have two and three speaker symbols. The symbol does not indicate the type of speaker. Letter designations for speakers are usually SPKR, SP, S and LS (LS means loudspeaker).

TAPE RECORDER HEADS

A tape recorder head is used to convert electrical signals from a microphone or other input into a magnetic field which varies with the input signal. The magnetic field in turn magnetizes the tape as it passes across the head. Schematic symbols for tape heads are shown in Fig. 2-16.

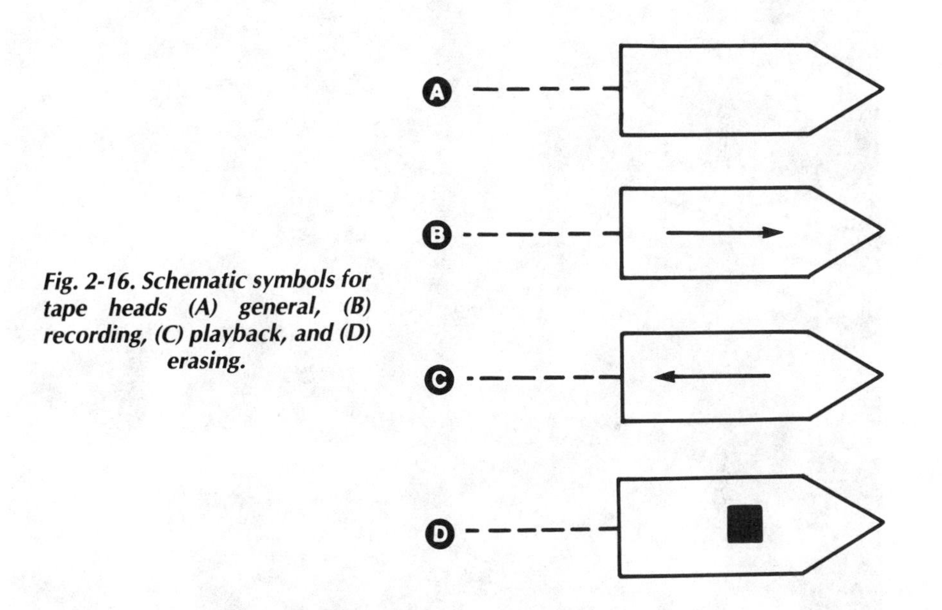

Fig. 2-16. Schematic symbols for tape heads (A) general, (B) recording, (C) playback, and (D) erasing.

Just as a recording tape head magnetizes the tape in accord with the input signal, a playback head picks up the pre-recorded variations from the tape and sends them to an amplifier for reproduction into sound. The schematic symbols for these heads are the same, but the letter designation indicates the function. For example, a playback head will be designated P, while a record head will be indicated R. In some cases, the same head is designed to provide both functions and it will be designated R/P. A tape recorder also uses a third head called the *erase* head. Its purpose is to erase any previously recorded material on the tape before it reaches the recording head.

The erase head, located before the recording head in the tape path, provides a high-frequency magnetic field which alternately increases and decreases to eliminate previous tape magnetization. The erase head is indicated by the same symbol used for record and playback, but the letter designation is the letter E.

HEATSINKS

Heatsinks are used to remove excess heat from such components as power transistors, heavy-duty resistors, and some types of vacuum tubes. Heatsinks usually have fins that radiate the excess heat into the surrounding air. The high-quality heatsinks shown in Figs. 2-17 through 2-19 are manufactured by Thermalloy, Inc.

Fig. 2-17. These are omni-directional heatsinks for LSI devices and gate arrays (courtesy of Thermalloy, Inc.).

Fig. 2-18. These high-performance TO-3 heatsinks are used on pc boards where there is limited space available (courtesy of Thermalloy, Inc.).

Fig. 2-19. This is an extruded heatsink for a plastic case or high-power applications (courtesy of Thermalloy, Inc.).

EXPERIMENTER'S BREADBOARD

One of the most useful miscellaneous "components" for the hobbyist and experimenter is the solderless breadboard. Both the electronics components and the interconnecting wiring plug directly into the solderless connectors on the board. The holes in the board are arranged in rows and columns in such a way that five holes in each row are connected together electrically, while the columns are electrically separated. Special rows on the sides are used for circuit ground connections. These breadboards make it very easy to experiment with different circuit configurations. Figure 2-20 shows an experimenter's breadboard.

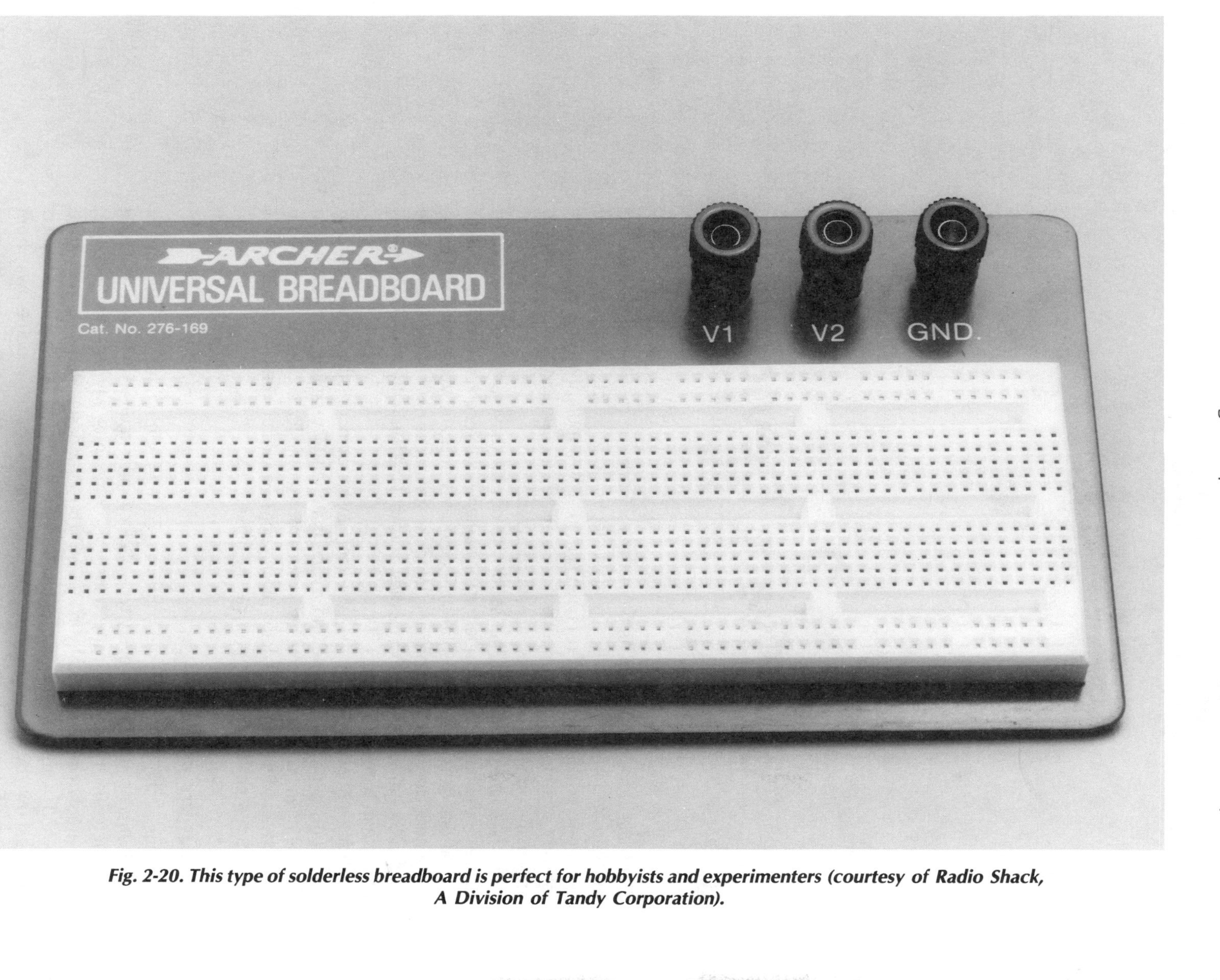

Fig. 2-20. This type of solderless breadboard is perfect for hobbyists and experimenters (courtesy of Radio Shack, A Division of Tandy Corporation).

3
Solid-State Devices

Solid-state devices, or *semiconductors* as they are commonly called, are used today instead of vacuum tubes in most electronic products. There are several reasons why solid-state devices have gained such popularity in the electronics industry. Among the most important are that they require less power to operate, are smaller in size, cost less, and last longer.

It is an interesting fact that solid-state devices have only recently gained tremendous popularity, considering the earliest radio receivers used cat-whisker crystal detectors that were solid-state devices. For many years, obviously, the semiconductor field was dormant. With the advances now being made in the state-of-the-art, solid-state devices, they have replaced virtually all of the functions of vacuum tubes.

Semiconductors are made of either germanium or silicon that contain impurities to produce *n-type* and *p-type* materials. The actual physics behind the operation of solid-state devices is much more complicated, but for the purpose of understanding semiconductor placement in a circuit, the p-type material is considered positive and the n-type, negative.

TRANSISTORS

One of the most common solid-state devices is the transistor. Because of its advantages, it is used in every type of electronic circuit with more uses being discovered every day. Basically, transistors come in two types—npn and pnp. These are shown schematically in Fig. 3-1. The letter designations normally used are Q or T. Other variations of transistors are discussed later in this chapter.

As shown by the symbols, the npn transistor emitter arrow points away from the base junction while the pnp arrow points toward the base. This is the way all pn and np junctions are designated. Remember these symbols when working with solid-state circuits, because the *bias voltages* (the voltage that causes a semiconductor to pass current) are opposite for the pnp and npn types, and the two cannot be interchanged. In other words, a pnp transistor must be replaced with a pnp.

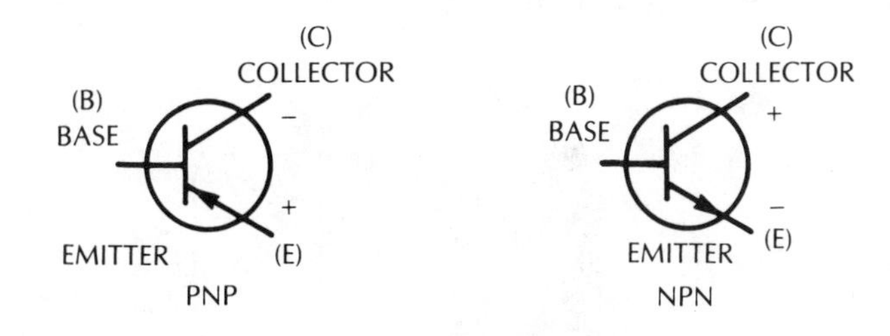

Fig. 3-1. Transistor Symbols show whether they are npn or pnp.

Transistorized circuits are similar to basic tube circuits except they are physically much smaller and use lower voltages. Simplified schematics of a transistorized i-f stage and an equivalent tube circuit are shown in Figs. 3-2 and 3-3.

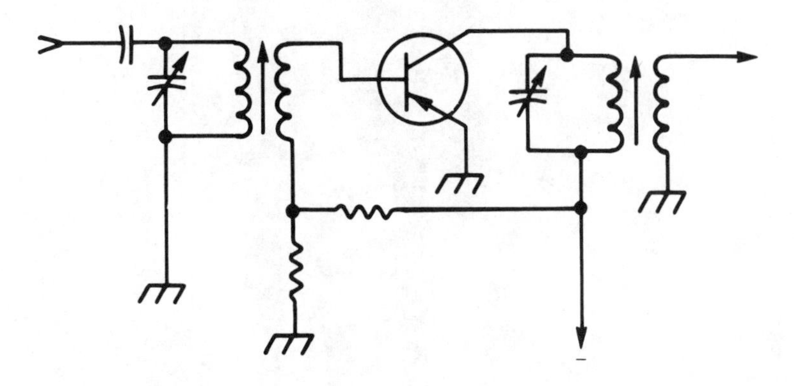

Fig. 3-2. A transistorized i-f stage closely resembles the tube-type i-f stage shown in Fig. 3-3.

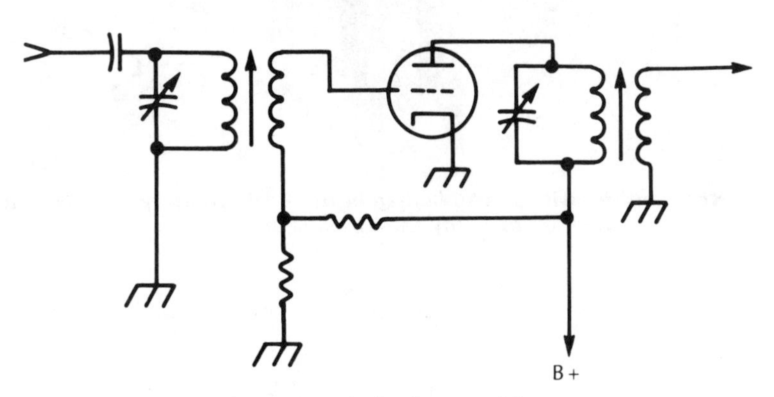

Fig. 3-3. Typical tube-type i-f stage.

Typically, transistors used in low-power circuits are mounted directly on their own leads and have the appearance of the one shown in Fig. 3-4. High-power transistors are physically larger with heavier leads to handle the larger currents (see Fig. 3-5). Transistors of this type are often mounted on a heat sink, which again is a metal mounting surface capable of dissipating a greater amount of heat than the transistor case itself.

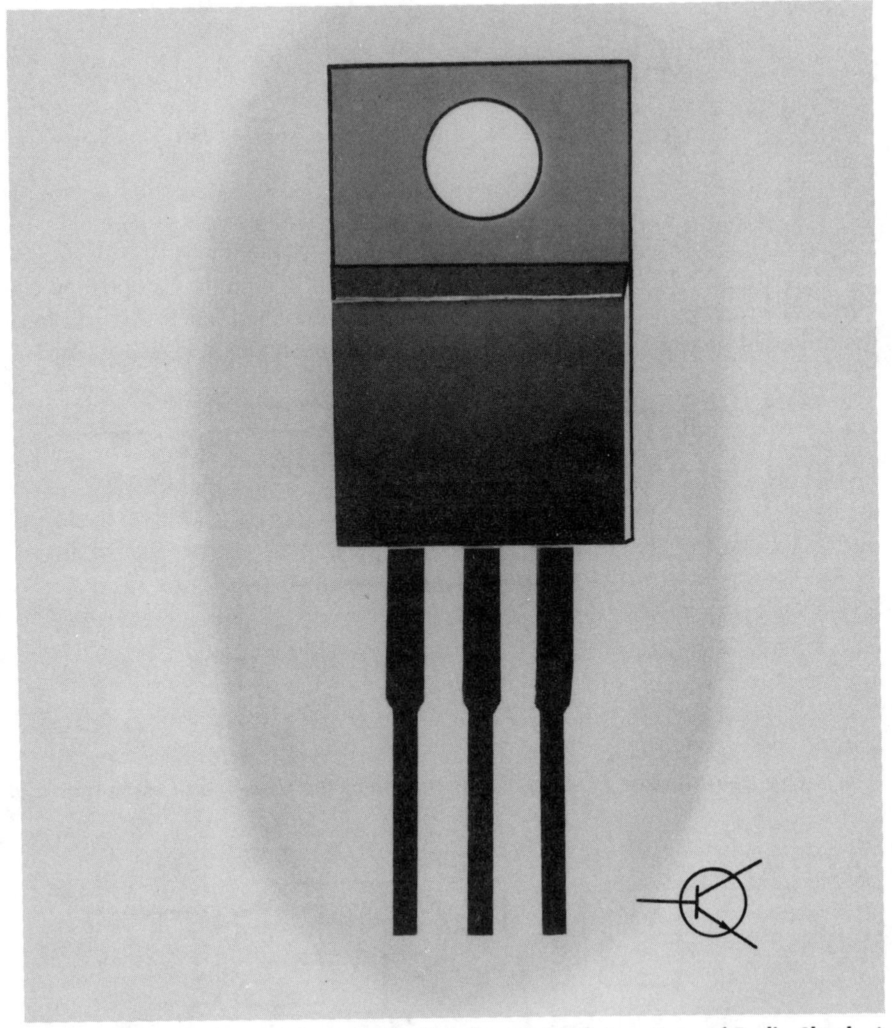

Fig. 3-4. Npn transistor with a small built-in heatsink tab (courtesy of Radio Shack, A Division of Tandy Corporation) and symbol.

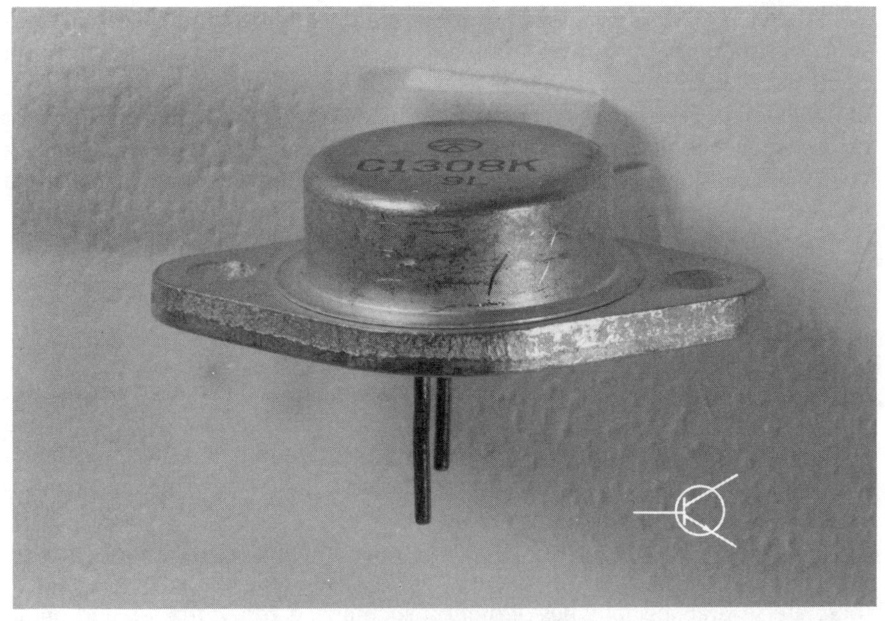

Fig. 3-5. A power transistor in a TO-3 case (courtesy of Radio Shack, A Division of Tandy Corporation) with symbol.

The physical appearance and size of a transistor is often referred to as its *package type* or *case configuration*. Packages have been designated with a standardized group of letters and numbers to make physical identification easier, for example, TO-3, TO-16, TO-32. The transistor shown in Fig. 3-5 is in a TO-3 package. Figure 3-6 illustrates another basic type.

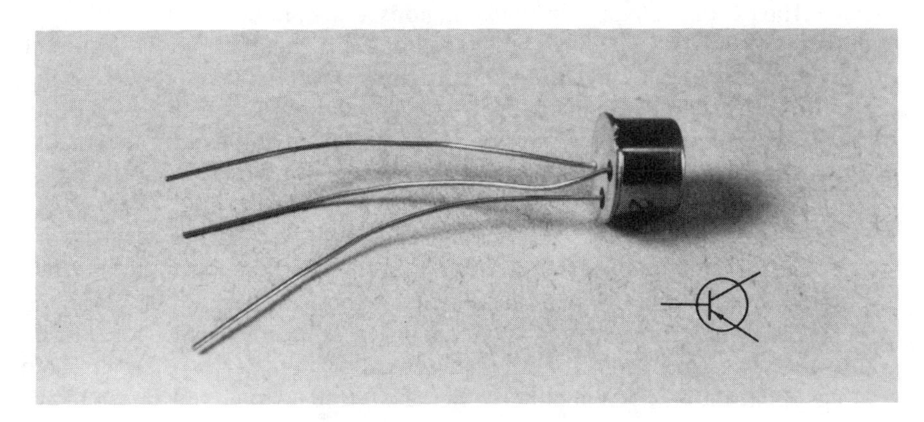

Fig. 3-6. This low-power "top hat" transistor is a germanium type.

The types of transistors used in various circuit applications also vary in construction, although they are still pnp or npn types. Transistor construction can be of either germanium or silicon and it can be a junction, epitaxial, mesa, unijunction, or field-effect type. Each has its own operating characteristics for certain applications. Generally, a junction transistor is used as an amplifier. An epitaxial transistor is better for higher power and frequency operation. A mesa transistor is somewhat like the junction but provides better high-frequency gain, is more rugged, and will handle more power. Finally, the unijunction and field-effect devices are discussed later under the heading SPECIAL DEVICES.

DIODES

Solid-state diodes can be described simply as being one-half of a transistor. In other words, a diode can be germanium or silicon with one piece of n-type and one piece of p-type material. The two pieces are joined, as in a transistor, at a point called a *junction*. Also, as in a transistor, the p-type material is considered positive and the n-type, negative.

The schematic symbol for a diode is shown in Fig. 3-7. The most common letter designation is CR; however, the letter D is also used. As might be expected, there are several special types of diodes, such as the zener, tunnel, and photo diodes which are discussed later in this chapter.

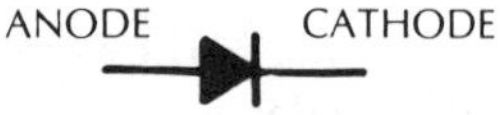

ANODE CATHODE

Fig. 3-7. Schematic symbol for a diode.

A solid-state diode functions like a tube diode where current passes freely in one direction but not the other. Therefore, a diode is a *unilateral* (one-way) device. The p-type material (positive) functions as the anode and the n-type, the cathode. A positive bias voltage on the cathode will therefore not pass current through the diode, but the diode will conduct if positive bias is applied to the anode (see Fig. 3-7). A diode can be checked with an ohmmeter: one polarity (direction) should be infinite (or very high) resistance, and the opposite polarity should be zero (or very low). Hence, if the reading is low both ways, the diode is shorted.

The diode itself is marked to designate the cathode end. Polarity may be shown by an actual diode symbol printed on the body, or it can be a dot or a colored strip. Examples of various diodes are shown in Fig. 3-8.

Solid-state diodes, just as transistors, are made for specific applications. Those used in low-power circuits are physically smaller than devices used in large current applications. Some mount by their axial leads while others have heavy studs designed for heat dissipation.

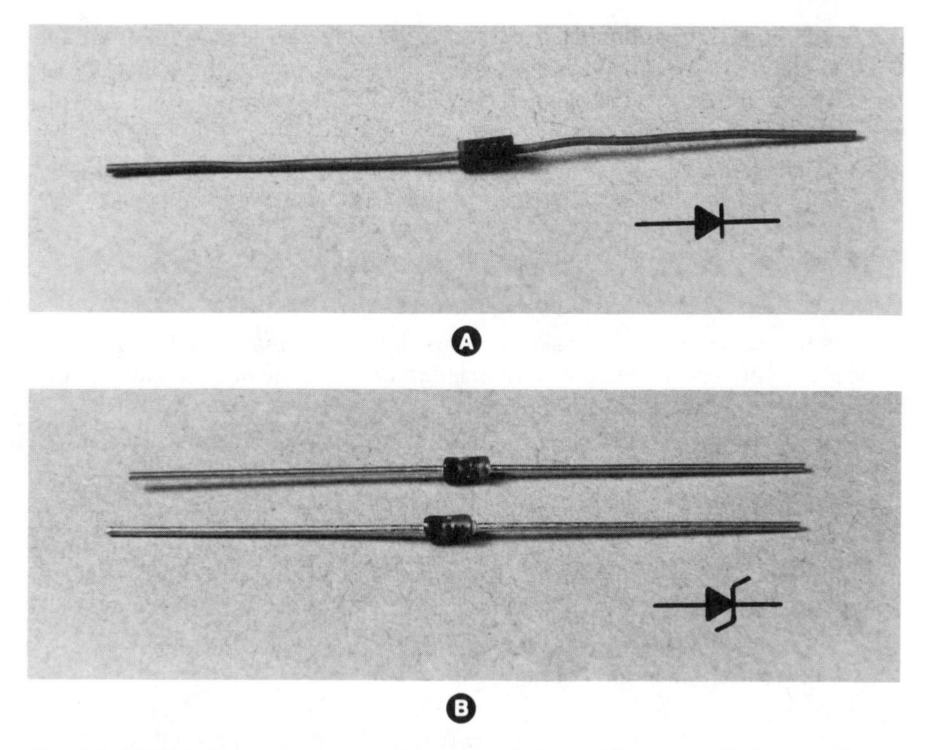

Fig. 3-8. (A) This is a typical general-purpose diode. (B) These two diodes are 12-volt zeners.

Zener Diodes

A *zener diode* is a special device used primarily for "holding" a voltage within a given limit. Zener diodes are also sometimes referred to as backward diodes or voltage-regulator diodes. A zener is similar in operation to a silicon junction diode, except it can also operate in *reverse bias* (opposite polarity of bias voltage). The zener maintains a fixed voltage because it conducts when the reverse bias reaches a specific value. In a zener regulating circuit, a series resistor is used to limit the diode current once it starts conducting, and the result of zener conduction is that it compensates for any increase or decrease in the power supply voltage (caused by a decrease or increase of load current).

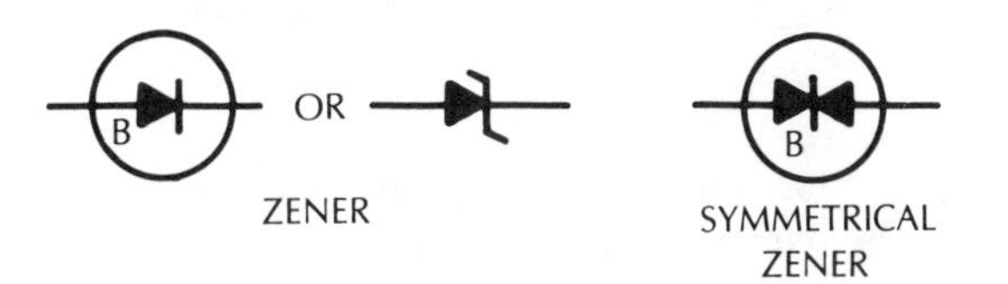

Fig. 3-9. Special symbols are used to indicate a zener diode.

Zener diodes are available for use in circuits with voltages ranging from 2.4 to 200 volts and with power ratings as high as 50 watts. On the schematic, a zener is normally designated by the letters ZD, ZR, or in some cases just D with the schematic symbol and parts list indicating that it is a zener. Because of the zener's small size, long life, and wide selection of available operating voltages, it has gained tremendous popularity in electronic equipment—especially when stability is necessary.

Tunnel Diodes

The tunnel diode (or the Esaki diode) is shown schematically in Fig. 3-10. It has some very special features that make it a popular device for low-power, high-frequency applications.

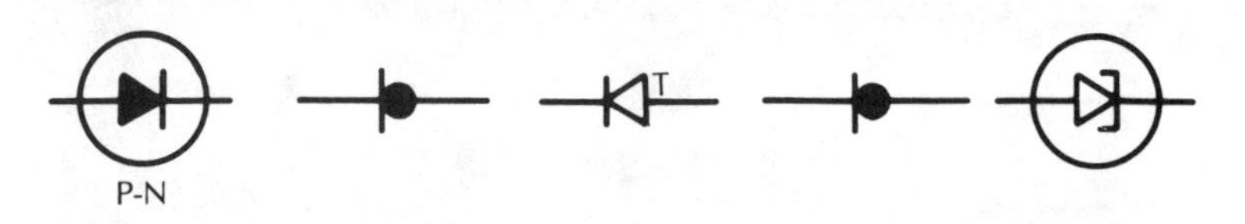

Fig. 3-10. These symbols are used to indicate a tunnel diode.

The name "tunnel" comes from the way electrical charges go through the junction barrier region on one side of the diode; they seem to disappear while another charge abruptly appears on the other side. This happens at nearly the speed of light. Some of the important features of this device are that it can operate at higher temperatures than silicon or germanium diodes and, furthermore, at frequencies as high as 10 GHz. Tunnel diodes can be made extremely small in size.

Varactor Diodes

Varactor diodes employ a condition that exists at a pn-junction diode when reverse bias is applied, a condition that makes the diode act like a capacitor. Schematic symbols for a varactor diode are shown in Fig. 3-11.

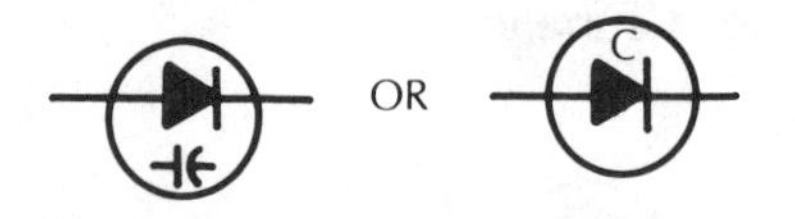

Fig. 3-11. Varactor diodes are represented by the symbols shown here.

Recall that a capacitor is made of two plates (conductors) separated by an insulating material. The *barrier region* (the region around the pn junction which is devoid of current carriers) of a diode acts just like a capacitor when it is reversed biased. The reason the pn junction acts like a capacitor is quite involved, but simply think of each side of the junction as the plates of a capacitor. As the bias voltage varies, the spacing between the plates varies and so does the capacitance. In other words, a changing dc voltage is used to vary the amount of conductance or "tune" the diode. Varactor diodes have recently become very popular in oscillator tuning circuits.

INTEGRATED CIRCUITS

Have you ever wondered what kind of devices it takes to build a pocket-sized TV set, a receiver smaller than a postage stamp, or a complete audio amplifier in less space than a period at the end of a sentence? Obviously it calls for miniature components, or more specifically, integrated circuits. An integrated circuit—or IC—is a single "component" containing resistors, capacitors, diodes, and transistors on one, tiny chip. The block diagram in Fig. 3-12 illustrates the large number of stages that can be built into just one of these devices. Figure 3-13 shows the dual in-line package (a chip with two straight rows of leads).

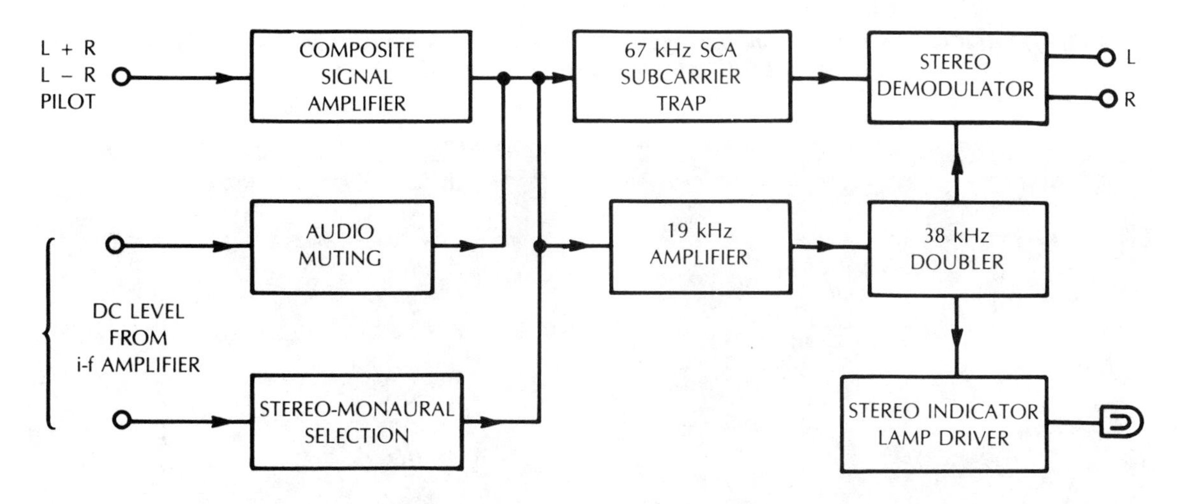

Fig. 3-12. Block diagram of an integrated-circuit stereo demodulator.

Integrated circuits, in addition to being microminiature in size, have high reliability characteristics and can be mass produced, resulting in lower cost. They are becoming increasingly popular in home entertainment products such as hi-fi amplifiers, FM radio, and television receivers. An example of an IC in a "top hat" case is shown in Fig. 3-13(B).

The manufacturing of an integrated circuit requires many steps. Briefly, an IC starts as a thin slice of silicon. The circuit elements are built up on it as layers of doped semiconductor material and silicon dioxide. It is then completed by adding the connecting leads. Manufacturing processes are continually being perfected to reduce production costs and increase IC popularity.

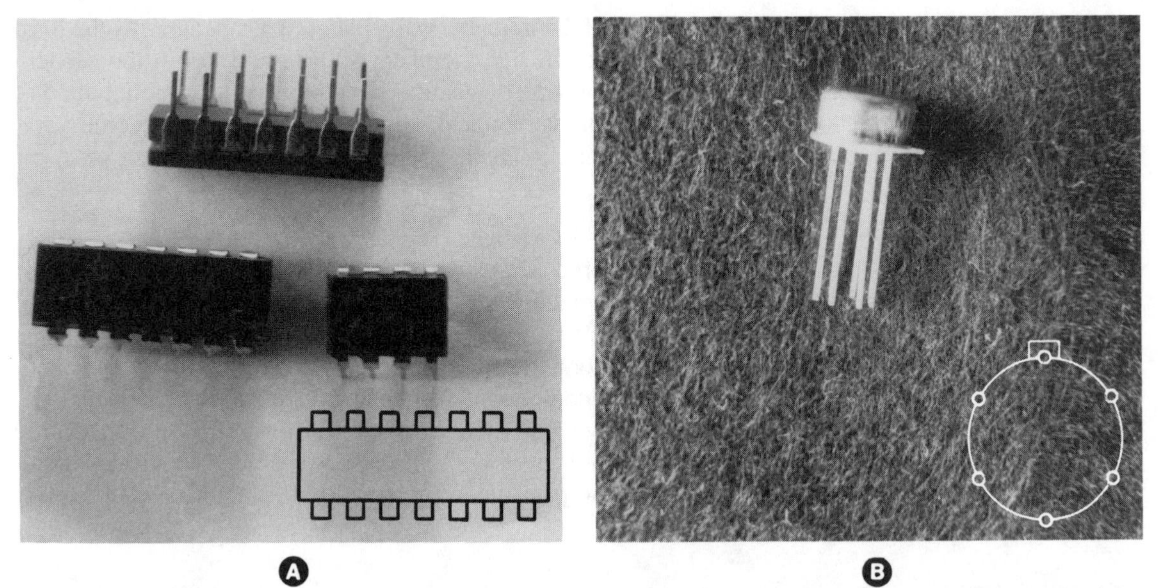

Fig. 3-13. A) Most ICs come in dual in-line packages, symbol depicting. Some ICs come in metal cans that look like transistors with extra leads, but the symbol is the same (B).

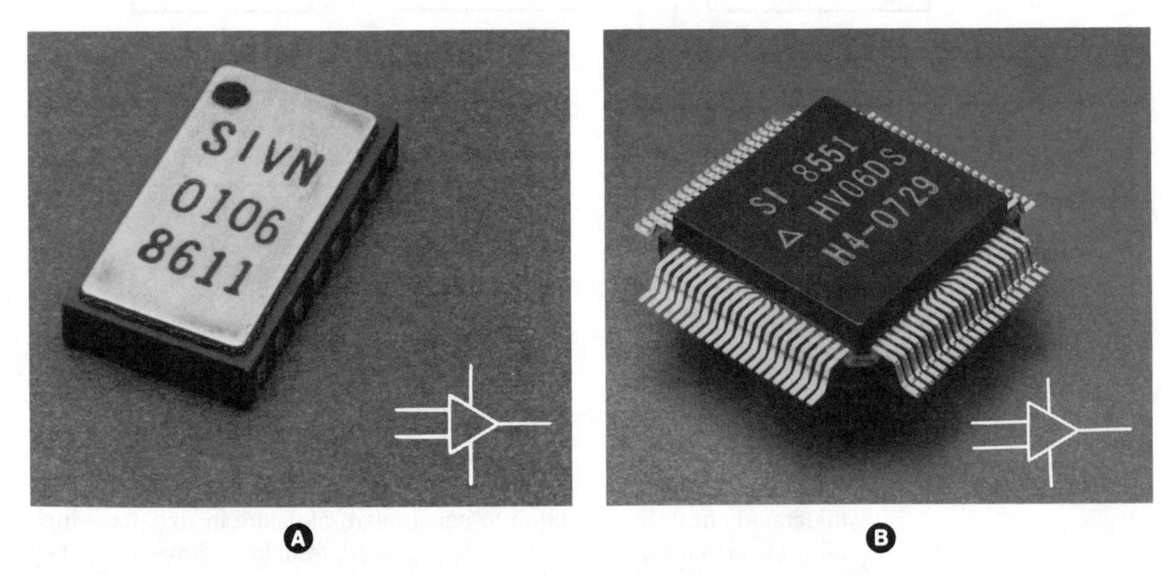

Fig. 3-14. (A) This is another style of dual in-line package. (B) Modern ICs can be made in packages with leads on all four sides (both photographs are courtesy of Supertex, Inc.). These ICs are operational amplifiers, as the symbol indicates.

Integrated circuits are normally soldered into a circuit just like transistors, except there are many more leads. Packaging is one of the biggest considerations in IC technology. Many ICs currently used in consumer products are designed for ease of replacement since they themselves are not made to be repaired. Working on and around ICs requires the same careful attention given to any solid-state device.

SPECIAL DEVICES

Silicon-Controlled Rectifier (SCR). The silicon controlled rectifier is identified by the schematic symbol shown in Fig. 3-15. It is designed for high power applications such as rectifying (changing ac to dc), regulation, and switching in motor speed control and light dimming circuits, for example. An SCR can be compared to a switch because it acts somewhat like a diode. That is, when an SCR is operating, its impedance (or resistance) to current flow is very low. When an SCR is turned off (or not conducting), its impedance is very high.

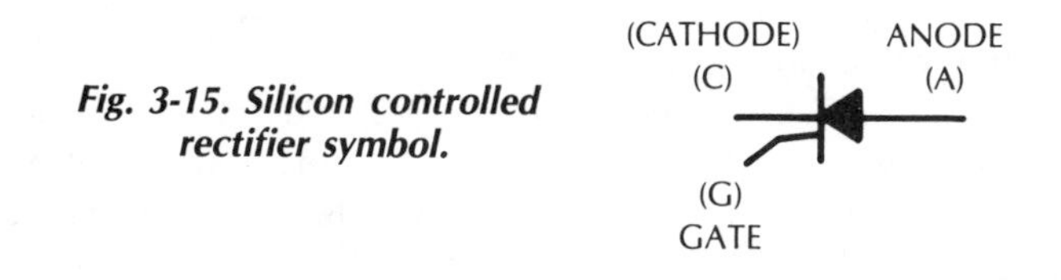

Fig. 3-15. Silicon controlled rectifier symbol.

As shown in Fig. 3-15, the SCR has three terminals labeled anode, cathode, and gate. When connected to a circuit, a positive signal applied to the *gate* terminal is used to switch the SCR on. Once it is switched on, it can be shut off only by lowering its anode voltage to a specific level. The anode and cathode operate the same as their counterparts in a normal diode. Like other members of the semiconductor family, the SCR comes in many shapes and sizes, depending on the circuit requirements.

Silicon-Controlled Switch (SCS). This device is similar to the SCR, except, as shown in the schematic symbol in Fig. 3-16, it has two gate terminals instead of one. The SCS operates at lower current levels and requires a negative gate signal to turn it on rather than a positive one.

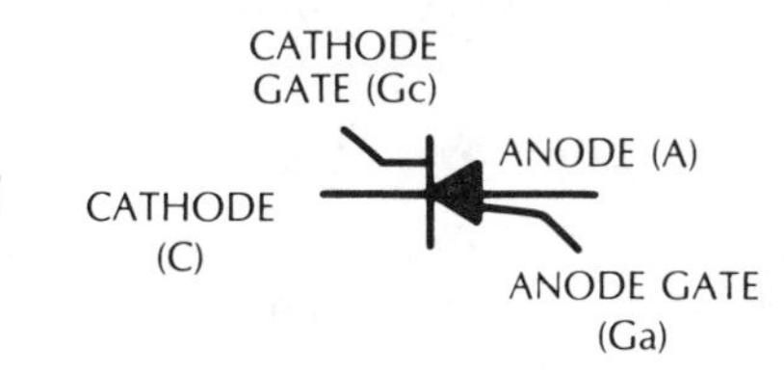

Fig. 3-16. Silicon controlled switch symbol.

Diac. The *diac* looks like two diodes in parallel as shown in the schematic symbol of Fig. 3-17. The name itself suggests a dual function that represents how this device works.

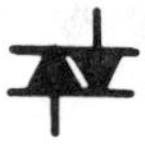

Fig. 3-17. Symbol for a diac.

The diac will conduct regardless of the polarity of the applied voltage. As voltage is applied, one-half of the diac is turned on as the proper level or *breakdown voltage* is exceeded. The other half is biased off. Then, if the input polarity is switched, the opposite half of the diac will conduct while the previously conducting half is turned off.

Diacs can be used as relaxation oscillators, but they tend to stay in conduction when the input signal voltage exceeds 50 Hz; so their effective operation is restricted to low-frequency applications.

Triac. A triac is a unique device resembling an SCR because it is used in controlled circuits, it can handle large current levels, and can be switched on with a gate. As the schematic symbol in Fig. 3-18 illustrates, the triac looks like a diac with the addition of a gate terminal. Furthermore, it operates on either voltage polarity like a diac. However, like the diac and SCR, it is effective only at low frequencies (approximately 50 Hz).

Fig. 3-18. The triac symbol shows the addition of a control element to a diac symbol.

In reference to diacs, it was noted that one-half of a diac conducts when its breakdown voltage is exceeded while the other side is biased off. This condition reverses when the polarity of the applied voltage reverses. A triac works the same way with one exception—its gate circuit allows it to be turned on, even if the applied voltage does not reach the breakdown voltage level. Therefore, the gate acts like a switch.

FIELD-EFFECT TRANSISTORS (FET)

The *field-effect transistor* is available in two types. These include the *junction field-effect transistor,* known simply as the field-effect transistor or FET, and the *insulated-gate, field-effect transistor,* or IGFET. The latter device is also referred to as the *metal-oxide semiconductor* or MOSFET. The IGFET (or MOSFET) is described later.

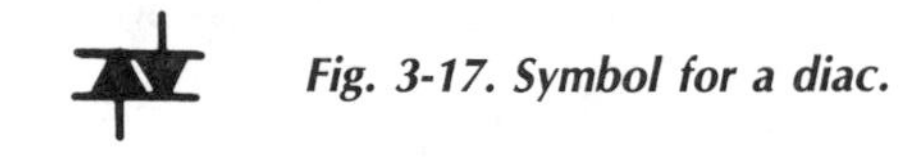

The FET is very similar in operation to a vacuum tube. Its main feature is that it is a very high-impedance device; therefore, it is popular for input circuits in voltmeters and other measuring instruments where minimum circuit loading is desired. An ordinary voltmeter, or one that is not a high-impedance type, can cause a change in the circuit being measured and result in inaccurate readings.

The schematic symbol for the FET is shown in Fig. 3-19. Like a transistor, it can be either npn or pnp. Its gate, drain, and source terminals are comparable to the grid, plate, and cathode of a vacuum tube. The source and drain terminals are connected to opposite ends of a section of n-type material. Current will flow from source to drain when the proper voltage is applied (and with no voltage applied between the gate and source). As in the grid circuit of a vacuum tube, when voltage is applied between the gate (grid) terminal and the source (cathode) terminal, it will regulate current flow.

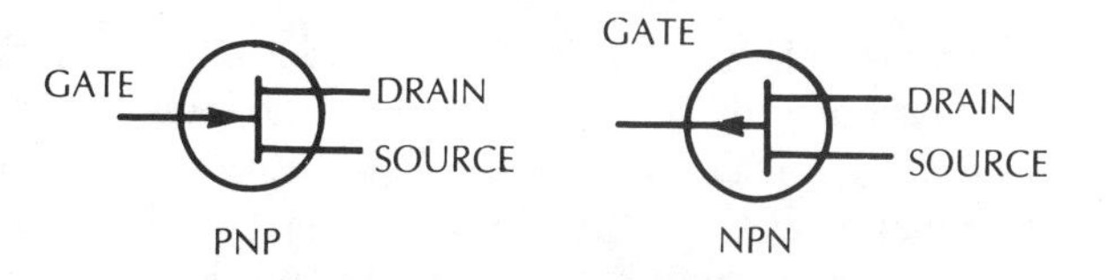

Fig. 3-19. FETs are represented by this symbol.

IGFETS AND MOSFETS

The insulated-gate, field-effect transistor (IGFET) is similar to the FET except its gate is insulated by a material called a *substrate*. As a result, its impedance characteristics are improved over the FET.

Like the FET, an IGFET can be either an npn- or pnp-type, as shown in Fig. 3-20. It is effective for input circuits of test instruments and audio applications where high impedance is desirable. The IGFET or MOSFET requires careful handling because of its high-impedance characteristic: the static electricity of a person's body is enough to damage this device. For this reason, manufacturers often caution users to handle IGFETs and MOSFETs by the case rather than by the leads.

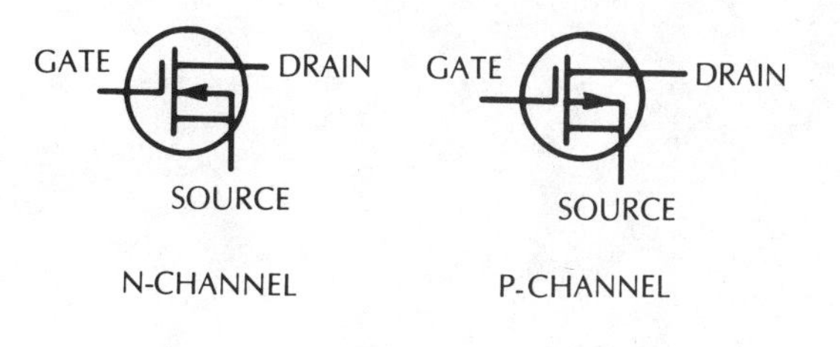

Fig. 3-20. Symbols for IGFETs and MOSFETs.

UNIJUNCTION TRANSISTOR (UJT)

The *unijunction transistor* is shown schematically in Fig. 3-21. It looks like the FET but actually operates more like a tunnel diode. The UJT is basically a low-frequency switching device used in timing circuits, voltage comparators, relaxation oscillators, and for triggering SCRs in control applications. Its physical appearance is like any transistor.

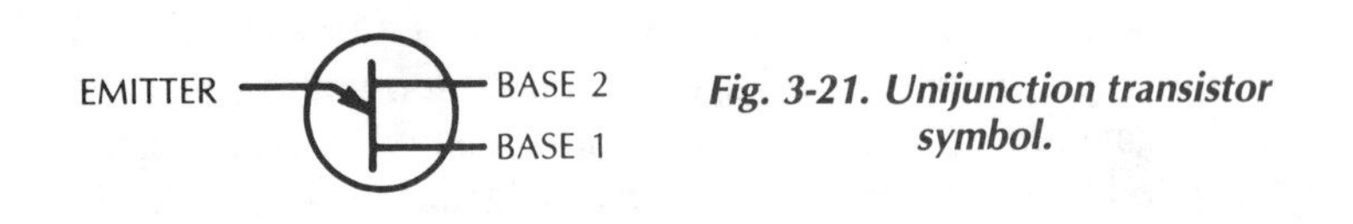

Fig. 3-21. Unijunction transistor symbol.

LEDS

Light-emitting diodes (LEDs) are special semiconductor devices that emit light when properly forward biased. The light emitted can be in either the visible or the infrared spectrum. An LED will not emit light if it is reverse biased.

LEDs are used mainly as indicating devices. They are usually used to show that a piece of equipment or particular channel is on. If the LED is dark, the equipment or channel is presumed to be off. A special type of LED is the seven-segment display. This device consists of seven separate LEDs and is used to display the numbers zero through nine and the letters A through F. An eighth LED is frequently used to display a decimal point. See Fig. 3-22. Figure 3-23 shows an LED and its schematic symbol.

Fig. 3-22. A seven-segment display LED and symbol.

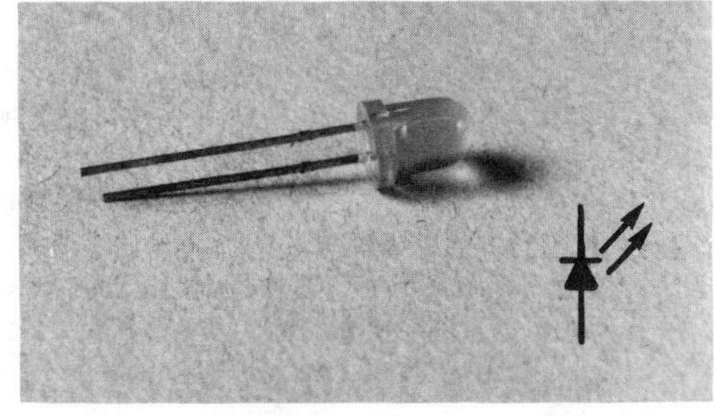

Fig. 3-23. A small, general-purpose LED and symbol.

LCDS

Liquid-crystal displays (LCDs) have replaced LED seven-segment displays for many applications because LCDs require only a very small amount of power. A scientific calculator, for example, using an LCD display, requires only 0.00024 watts. An equivalent calculator using LEDs would require hundreds of times as much power. The main disadvantage of LCD displays is that they are not easily visible in dim light.

An LCD is composed of two very thin sheets of metallic film with a special *nematic liquid* between the sheets. This nematic liquid contains molecules that align their long axes in parallel lines when excited by a very small electric current. This causes a particular segment of the metallic film to become dark. Standard seven-segment displays are now more often used with LCDs.

PHOTOCELLS

Photovoltaic cells, or *solar cells*, are semiconductor devices that produce an electric current when exposed to light. A photovoltaic cell is a pn junction that produces about 0.45 volts when exposed to sunlight. The amount of current that it is capable of producing depends upon the surface area of the cell. These cells can be connected in series and parallel to produce higher voltages and currents or both.

A different type of photocell is the cadmium-sulfide photocell. This device exhibits a large change in its resistance when it is exposed to light. Because of this phenomenon, cadmium-sulfide photocells are often used in light-actuated relays and switches. See Fig. 3-24.

Fig. 3-24. A cadmium-sulfide photocell. Note arrows in symbol.

INFRARED EMITTER AND DETECTOR

The *infrared emitter and detector* are two semiconductor devices that are often sold as a pair. The emitter consists of a special LED, usually made from *gallium-arsenide* that emits light in the near-infrared region of the light spectrum (780 to 1,000 nanometers). The detector is a special semiconductor device that is especially sensitive to the frequency of light produced by the emitter. The detector generates an electric current when exposed to the infrared energy.

Infrared emitters and detectors are often used for optical coupling between circuits that must be electrically separated. See Fig. 3-25.

Fig. 3-25. This is an infrared emitter and detector pair. The emitter is the clear LED on the left. Note symbols for respective components.

THERMISTORS

A *thermistor* (*therm*ally sensitive res*istor*) is a semiconductor device that exhibits a change in resistance when it is exposed to a change in temperature. On one hand, the resistance increases in a thermistor with a positive temperature coefficient. If the resistance decreases, however, the thermistor has a negative temperature coefficient. These devices are often used in thermal-protection circuits. If the measured temperature becomes too high or too low the thermistor will cause an alarm to sound or shut down the equipment.

4
Vacuum Tubes

Vacuum tubes are used in some types of electronic equipment and will continue to be for many years even though transistors and ICs have replaced them in most applications. However, there are still some functions and designs for which entirely suitable solid-state devices have yet to be perfected.

Tubes, of course, are available in a great many types and styles for various circuit functions, but they all have certain common characteristics. The electrodes, or elements of a tube, are housed in a glass or metal envelope that has been evacuated by drawing out all the gases. The most basic tube is the diode, so-called because it has only two elements—a cathode that emits electrons, and a plate, or anode, that collects the electrons. All tubes have at least these two elements.

The common schematic letter designation for tubes is the letter V, such as V2, V6, etc. However, dual tubes like the dual-diode or dual-triode, are usually indentified in two sections, such as V2A and V2B or V6A and V6B. (Further discussion on dual tubes occurs later.)

In some tubes, the cathode and filament are the same element. Such tubes have directly heated cathodes. Other tubes have indirectly heated cathodes in which a separate cathode is heated by a filament. The two types are shown in Fig. 4-1. The cathode, whether directly or indirectly heated, emits electrons into the vacuum. When the anode or plate is made positive with respect to the cathode, it will attract these electrons, resulting in current flow.

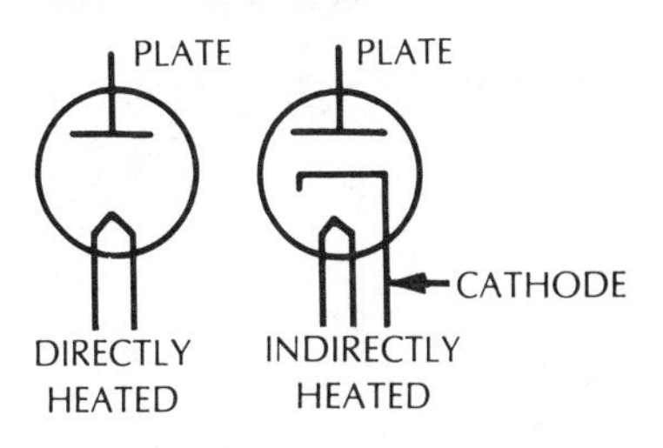

Fig. 4-1. All vacuum tubes have an electron emitter called the cathode. In some tubes, the heated filament is the emitter; in others the filament merely heats the cathode.

Because electrons must travel from the cathode to the plate, other elements can be introduced to control this current flow. Thus, a third element—called a *control grid*—is added and is capable of exerting significant influence over the electron stream from cathode to plate. Some tubes have more than one grid. The tube name indicates the number of elements. For example, the triode has three elements, the tetrode has four, and so on.

DIODES

The two-element tube is a *diode*. Its schematic symbol is shown in Fig. 4-2. The diode is most often used as a rectifier in power supply circuits. Because current flow in a diode will exist only when the plate is positive with respect to the cathode, an ac signal applied between the plate and cathode results in a current flow only during positive cycles.

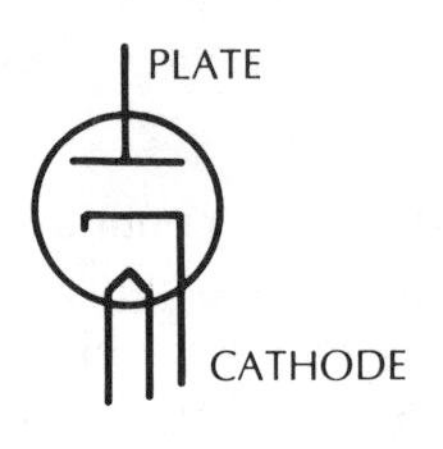

Fig. 4-2. Schematic symbol for a diode—a two-element tube.

A diode used in this application is termed a *half-wave rectifier* because it conducts on only one-half of the ac wave. A *full-wave rectifier* will provide output on both portions of the ac signal if two diodes are connected as shown in Fig. 4-3. A full-wave rectifier tube is actually two separate diodes in a single envelope. Both plates receive electrons from the same cathode.

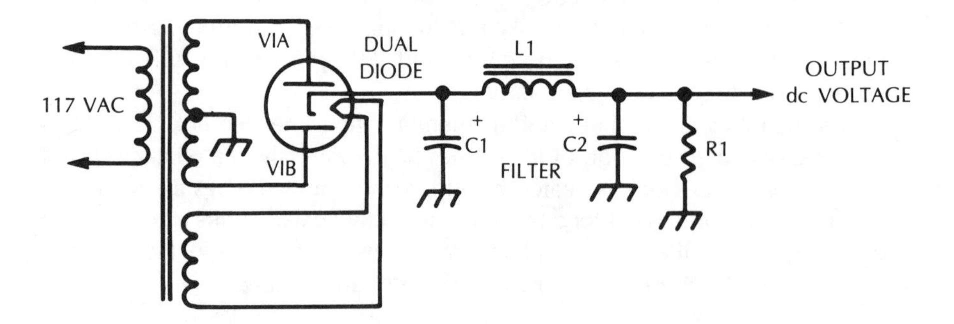

Fig. 4-3. Full-wave rectifier circuit using a dual-diode tube.

Another type of diode commonly used in power supply circuits is the gas-filled diode shown schematically in Fig. 4-4. One of the advantages of a gas diode is that it has more constant output and can handle a larger current flow with less power loss.

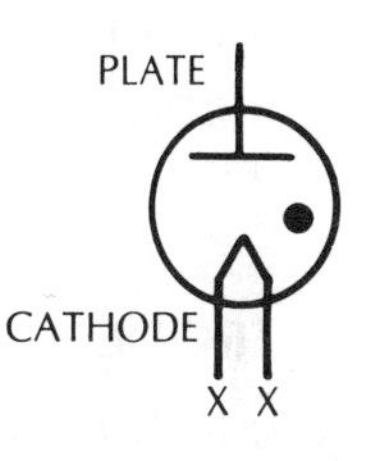

Fig. 4-4. A gas-filled diode is represented by the symbol shown here.

TRIODES

The name *triode* suggests a three-element tube—a cathode, a control grid, and a plate. The schematic symbol is shown in Fig. 4-5. The grid is placed between the cathode and plate to regulate the current flow through the tube; hence the name control grid. It acts much like a control valve in a water pipe. The grid is actually a fine mesh of wires surrounding the cathode and is closer to the cathode than the plate.

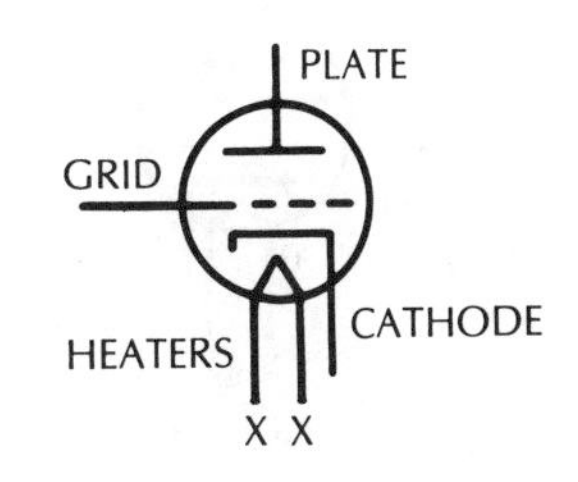

Fig. 4-5. By adding the third element to the diode symbol, we have the triode symbol.

In operation, electrons leave the heated cathode and are attracted to the plate (when the plate is positive with respect to the cathode). However, because some space exists between these two elements, not all of the electrons will reach the plate; some will collect around the cathode. Theoretically, the more positive the plate, the more electrons it will attract, but this is true only to a point. When the third element, the control grid, is placed close to the cathode and made slightly positive, it will essentially attract some of the electrons crowded around the grid to cause grid current. However, most electrons will be attracted by the more positive plate, resulting in a higher plate current. Because the grid is much closer to the cathode than the plate, even a small grid voltage will produce large effects on plate current. Hence, triodes are commonly used as amplifiers. As in the case of diodes, two triodes can be placed in one envelope; such a tube is known as a dual triode and is shown by the schematic symbol in Fig. 4-6.

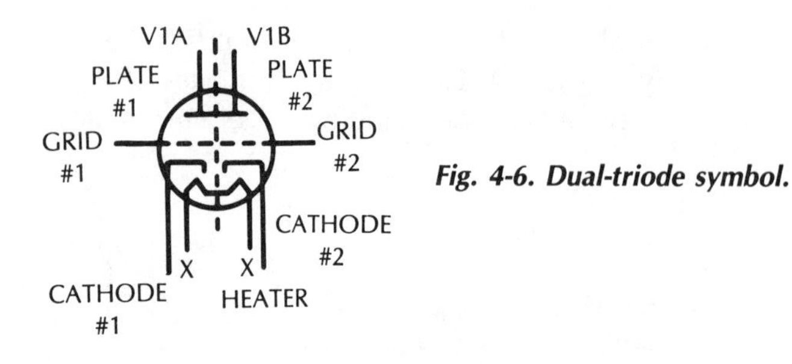

Fig. 4-6. Dual-triode symbol.

TETRODE

The *tetrode* is a four-element tube having two grids in addition to the two basic elements (plate and cathode). The tetrode schematic symbol, as illustrated in Fig. 4-7, shows the second grid—called a *screen* grid—between the control grid and the plate. The purpose of the screen grid is to reduce the capacitance between the control grid and the plate, a condition that can cause problems during high-frequency operation. The control grid still acts to regulate plate current as it does in the triode, but the control grid's effect is much greater in the tetrode, making it a higher-gain amplifier.

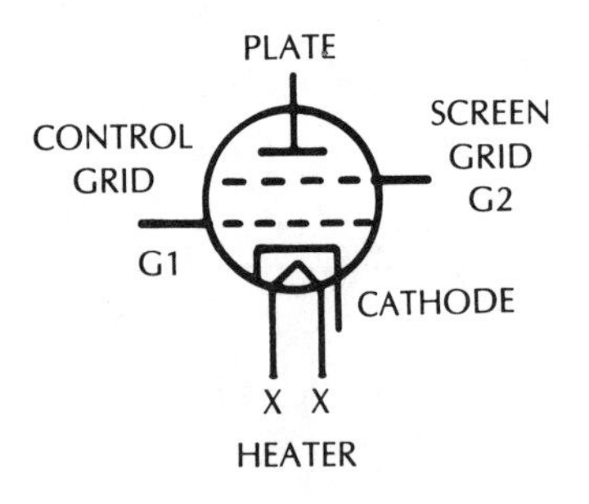

Fig. 4-7. The tetrode symbol shows the addition of a second grid between the first and plate.

In most circuit applications, the screen grid is bypassed to ground with a capacitor. This effectively places the screen grid at signal ground to further reduce the capacitance between the plate and control grid. The most common application for the tetrode is in amplifier circuits, however the pentode tube is even more widely used (except in certain applications).

PENTODE

The *pentode* tube has five elements—a cathode, plate, control grid, and screen grid, as in a tetrode—plus one additional element called the *suppressor* grid. The schematic symbol for the pentode is shown in Fig. 4-8.

In the tetrode, the screen grid was added to reduce the capacitance between the control grid and plate, resulting in better amplification at higher frequencies. However, during conduction, electrons hitting the plate cause some electrons to be "knocked off," producing an effect known as *secondary emission*. This secondary electron flow occurs when the electrons knocked off the plate are attracted by the screen grid which results in distortion and unwanted oscillations. To eliminate this effect, a suppressor grid was placed between the screen grid and plate. As shown in the pentode symbol (Fig. 4-8), the suppressor grid is tied to the cathode. Therefore, the suppressor grid is negative with respect to the plate, and as a result, repels secondary emissions from the plate.

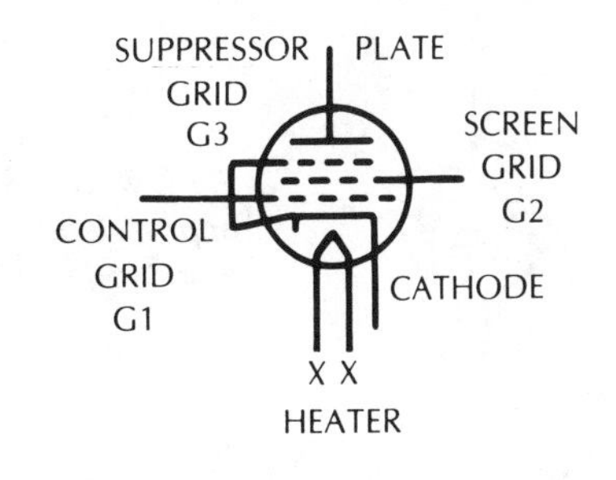

Fig. 4-8. Add still another grid to the tetrode symbol and it becomes a pentode.

The pentode has a much higher amplification factor than either the triode or tetrode and is used in amplifier stages where high gain and high plate resistance is desirable.

BEAM POWER TUBES

A *beam power tube* can be either a tetrode or pentode. The *beam power* characteristic comes from the fact that the electron stream from the cathode to the plate is formed into a narrow beam by special plates. In effect, these plates concentrate the flow of electrons much like funneling the water from a large pipe through a narrow nozzle, providing much greater pressure or force.

This concentration, or beam, does two things: it allows the tube to produce higher output power, and it reduces secondary emission. Greater power results because more electrons reach the plate faster than in an ordinary tetrode or pentode. Secondary emission, as mentioned earlier, occurs when electrons hit the plate and knock other electrons off. Because the electrons in a beam power tube are so concentrated, those that are knocked off the plate are immediately returned to the plate. The stray electrons would be moving against too heavy a current, so secondary emission is kept to a minimum. The schematic symbols for beam power tubes are shown in Fig. 4-9.

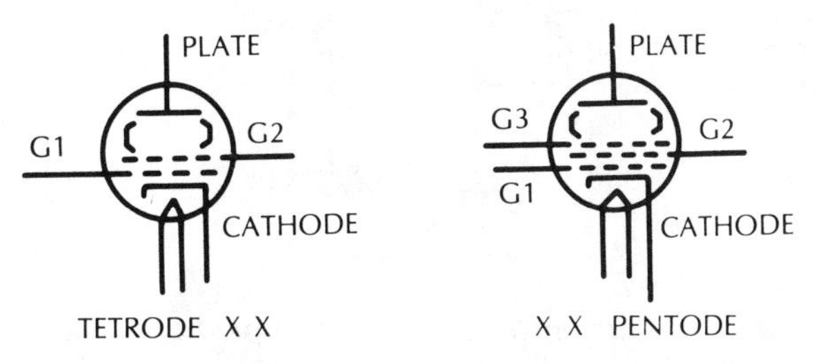

Fig. 4-9. These symbols represent beam power tubes.

PENTAGRID TUBES

As the name suggests, the *pentagrid tube* has five grids, as shown by the schematic symbol in Fig. 4-10. It is considered a multi-unit type, because in actual use it combines the functions of more than one tube. For instance, the pentagrid is often used as an oscillator, and/or mixer in superheterodyne receiver circuits. A pentagrid tube serving these functions is commonly referred to as a *pentagrid converter*.

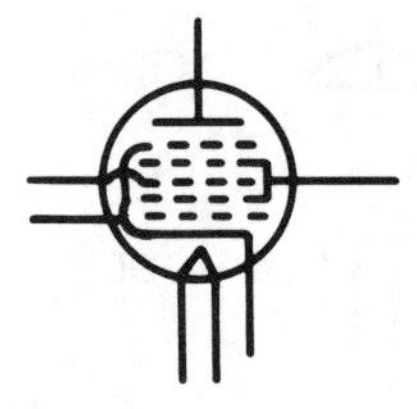

Fig. 4-10. A five-grid tube, appropriately called a pentagrid, was designed for converter circuits.

A simplified schematic of a typical oscillator mixer circuit using the pentagrid converter is shown in Fig. 4-11. Pentagrid converter circuits are common in ac, dc, and battery-operated units, including automobile broadcast receivers. They are also used as a *product detector* (when two signals are simultaneously applied to a detector circuit) in single side-band receivers.

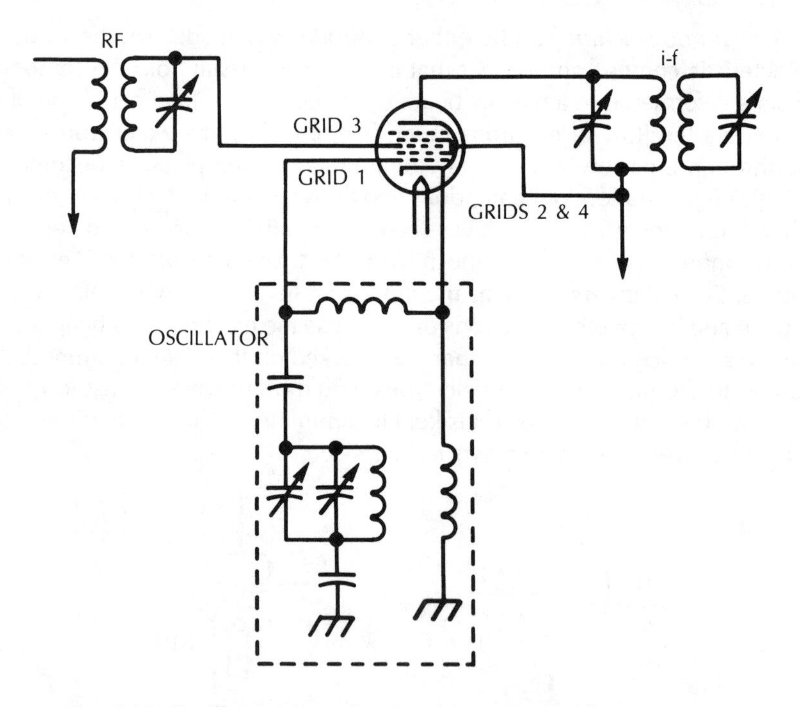

Fig. 4-11. Oscillator-mixer circuit using a pentagrid converter.

REGULATOR TUBES

Regulator tubes are used to maintain a constant power supply voltage regardless of load variations. In this respect, they serve the same purpose as zener diodes (discussed earlier). The regulator is a gas-filled tube that can regulate voltages, for example, at 75, 90, 105, or 150 volts. The gaseous regulator tube, (sometimes referred to as the "VR tube") is shown schematically in Fig. 4-12 and is designated by the letters VR.

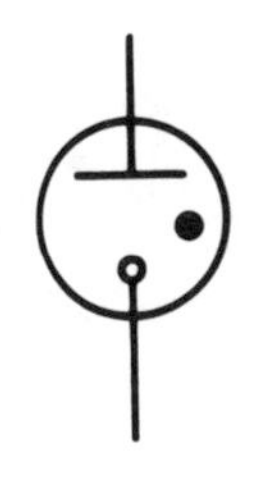

Fig. 4-12. Schematic symbol used for gaseous regulator tubes.

Typical circuit applications are illustrated in Figs. 4-13 and 4-14. The gaseous regulator in these circuits has a series limiting resistor just as that used with a zener diode. The resistor value is selected to hold the tube's current flow between specific levels. For most regulators, the minimum and maximum current values are between 5 and 40 mA. The voltage regulation capability of a gaseous tube is fairly good as long as the load current does not vary more than 30 mA. In other words, if the normal load current is 75 mA, the most it could vary to maintain proper regulation is from 60 to 90 mA, (± 15 mA or 30 mA total). The regulated voltage output should therefore be present across the tube.

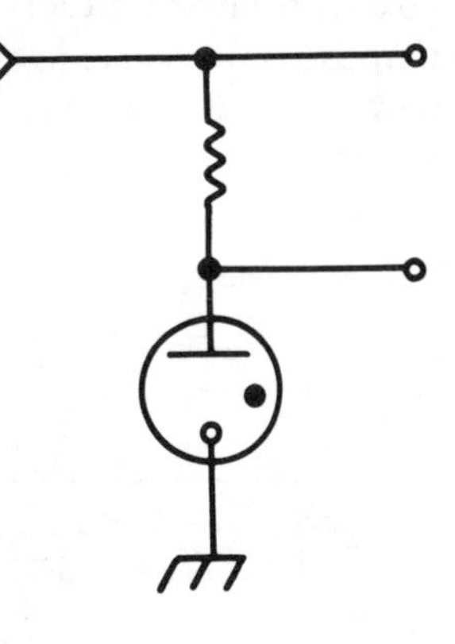

Fig. 4-13. Regulator circuit using a single VR tube.

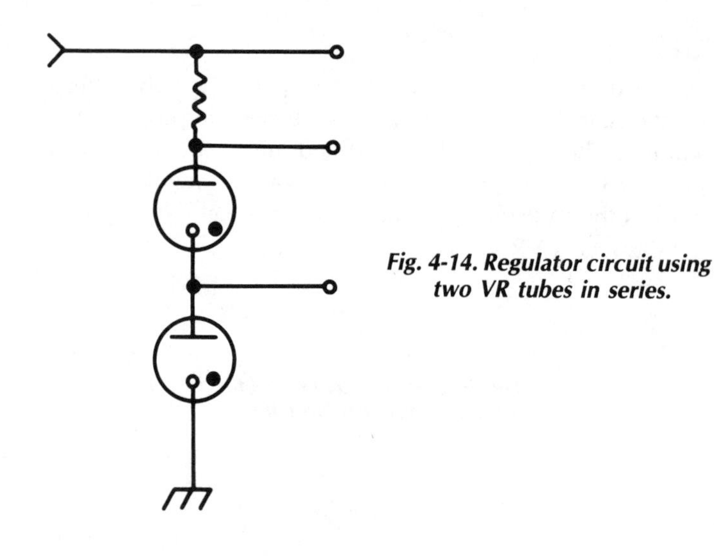

Fig. 4-14. Regulator circuit using two VR tubes in series.

CATHODE-RAY TUBES

The *cathode-ray* is the most familiar vacuum tube, because that's exactly what a TV picture tube is. The TV picture we see is actually formed by a beam of electrons swept back and forth (deflected) across the screen to "paint" an image. Naturally, there is a lot more involved than that, but the basic principle is the same for any cathode-ray tube (CRT).

CRTs come in two basic types: with electrostatic deflection or magnetic deflection. Both are shown in Figs. 4-15 and 4-16. As indicated, the electrostatic type has deflection plates within the tube itself, while an external deflection *yoke* is needed for CRTs designed for magnetic deflection. In both units, the purpose is the same: to sweep the electron beam from the tube gun back and forth across the screen, in step with the input signals, to produce the correct image.

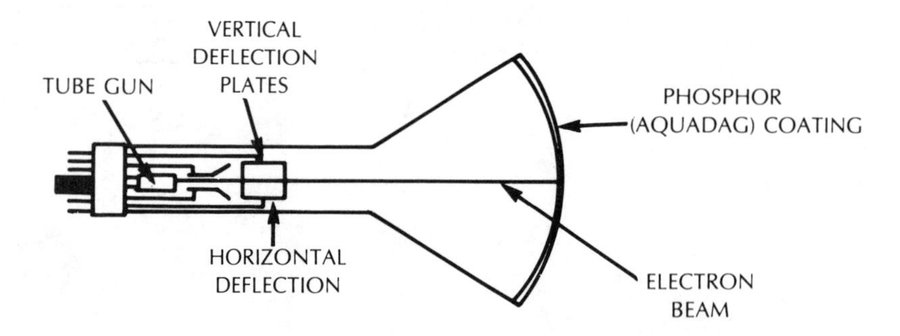

Fig. 4-15. Symbol for a cathode-ray tube designed for electrostatic deflection.

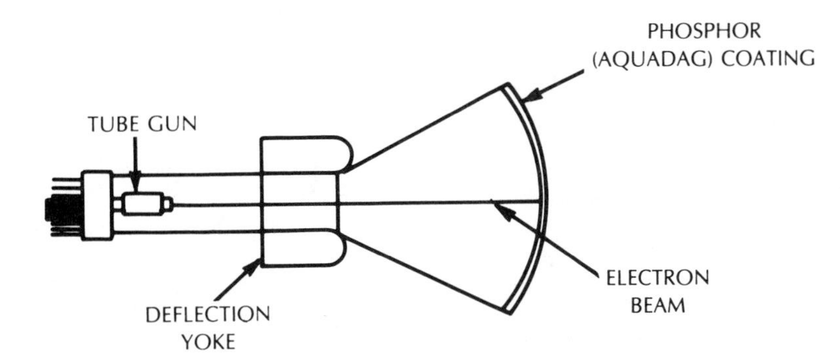

Fig. 4-16. Without internal plates, this symbol depicts a CRT requiring an external magnetic yoke.

Both CRT types are found in television sets, oscilloscope test instruments, and radar units. However, the magnetic-deflection type cathode-ray tubes are more commonly used in today's home television receiver. The schematic symbol for a cathode-ray tube is much like any other tube. A CRT has a filament, cathode, grids, and plate, as shown in Fig. 4-17. However, the plate is the phosphor coating on the inside of the tube faceplate and the voltage is on the order of thousands of volts compared to the few hundred volts in other types of tubes. The high voltage attracts the electron beam and speeds it past the anode to the phosphor-coated face, which glows under bombardment by the electron beam.

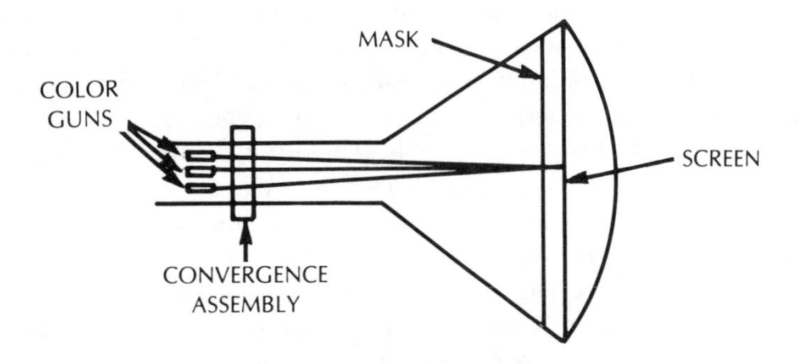

Fig. 4-17. Schematic symbol of a three-gun color CRT.

Color CRTs operate like the black-and-white type in that an electron beam is swept back and forth in synchronzation with input signals to form images. However, a color TV tube has three guns instead of one and the picture tube voltage is twice as high. The three guns are controlled by the primary colors—red, green, and blue—and produce three separate electron beams adjusted to pass through a special mask on the inside of the tube face. When correctly adjusted, the three electron beams converge at the same spot on the screen to produce a white dot. Hence, varying levels of each primary color are mixed to produce the spectrum needed to reproduce transmitted color signals.

5
Interconnecting Devices

Having developed an idea of the many types of components that go into making a radio, TV, or computer, the next question is, how are they all connected to form a circuit? In some electronic equipment, mostly large devices and older units, terminal strips, tube sockets, and plugs function as circuit connection points. Components are then soldered to the various terminal strips, sockets, etc., and wires are used to tie the required circuit points together.

Another, more compact technique is to use printed-circuit boards, on which thin layers of metal foil serve as interconnecting wires and tie points. In small transistor radios, for example, everything—including the battery, speaker, tuning control, on-off switch, and volume control—is mounted to the circuit board. Very little actual wire is used. However, in most electronic equipment, some type of interconnecting device is needed from one point to another or from unit to unit. For example, if a microphone is to be plugged into a stereo amplifier, a special cable and connector are needed to couple the voice signals to the audio input circuit. In larger equipment, wiring or cabling is necessary to connect various components together.

PRINTED-CIRCUIT BOARDS

A printed-circuit board is shown in Fig. 5-1. The components are all connected by foil patterns, made up of "lands" and "traces" that are etched onto the board, usually with a thin coating of copper. Then holes are drilled through the board to allow component lead connections. Components are located on the top of the board (opposite the foil side), and their leads are placed through the holes to be soldered to the foil on the bottom of the board.

Fig. 5-1. A printed-circuit board (courtesy of Hewlett Packard).

Terminals or connectors are conveniently soldered to the circuit board to provide a means of wiring external parts into the circuit such as panel-mounted controls, speakers, etc. In more complicated printed circuits where many connections are required, a special type of connector is used. In such cases all necessary connecting points are brought to one area on the board and a special multi-contact connector is soldered to this area.

Another type of printed-circuit board is shown in Fig. 5-2. An industrial application requiring multiple circuit boards with plug-in type connectors is shown in Fig. 5-3. Printed-circuit boards are popular in hundreds of consumer and industrial electronics applications.

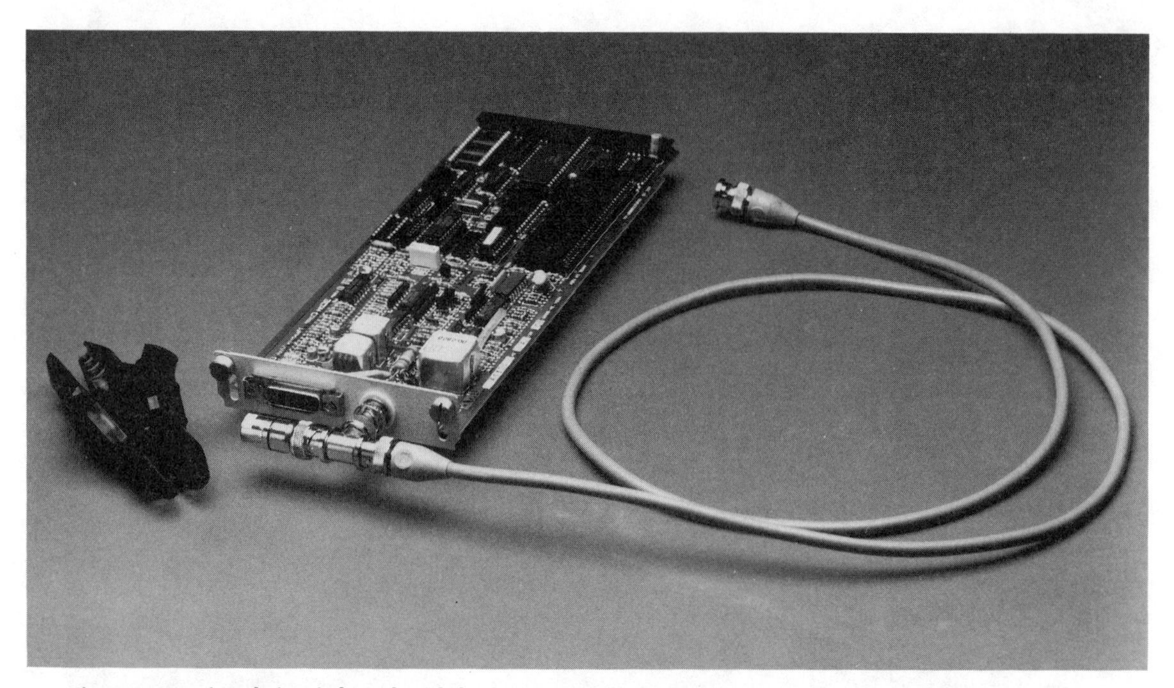

Fig. 5-2. A printed-circuit board with both large and small ICs (courtesy of Hewlett Packard).

Fig. 5-3. This is a small plug-in pc board.

WIRING HARNESSES

Wiring harness is the common name given a bundle of wires routed together from one circuit or piece of equipment to another. As an example, the wires from a TV deflection yoke are tied together to form a harness. The need for wiring harnesses is important, mandating neat, safe, economical procedures.

Wiring harnesses are needed in a number of applications such as from radio to radio in aircraft, between circuit board assemblies in a computer, from a common power supply to a receiver and transmitter, in a broadcast station from console to transmitter, and many others. In many instances, ready-made harnesses can be purchased with the required number of wires. These are called *cables* and come in hundreds of variations. The cabling in a telephone installation, for example, may contain as many as 100 wires.

The schematic symbol for multiconductor cable is shown in Fig. 5-4. In some applications, the entire harness or cable is enclosed in a metal or shielded jacket. This *shielding* electrically protects the information traveling through the cable from parasitic distortion by *shunting,* or shorting, the extraneous noise to ground.

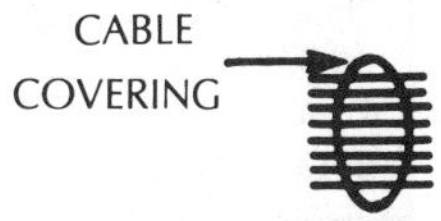

Fig. 5-4. Symbol used to signify unshielded multiconductor cable.

CABLE COVERING

The symbol for shielded cable is shown in Fig. 5-5. Multiconductor cables contain color-coded wires to make identification easier when connections are made. The wires may all be the same size or there may be several small and several large diameters for special applications requiring different voltages and currents.

Fig. 5-5. This symbol indicates shielded multiconductor cable.

Figure 5-6 is a drawing representing a bundle of wires or multiconductor cable connecting two circuits or units. Wire identification or coding is necessary for interconnecting cables, as indicated, and the coding is also denoted on the schematic.

MULTIPLE-WIRE CONNECTORS

Multiple-wire connectors are used to facilitate electrical connection between circuits and equipment. These connectors may be of almost any type or size depending on the application and number of wires. The connector may be a simple three-wire cable-to-cable type. Or it may be a special connector, designed to handle 100 wires of various sizes, with mechanical locking features to prevent the connectors from accidental separation.

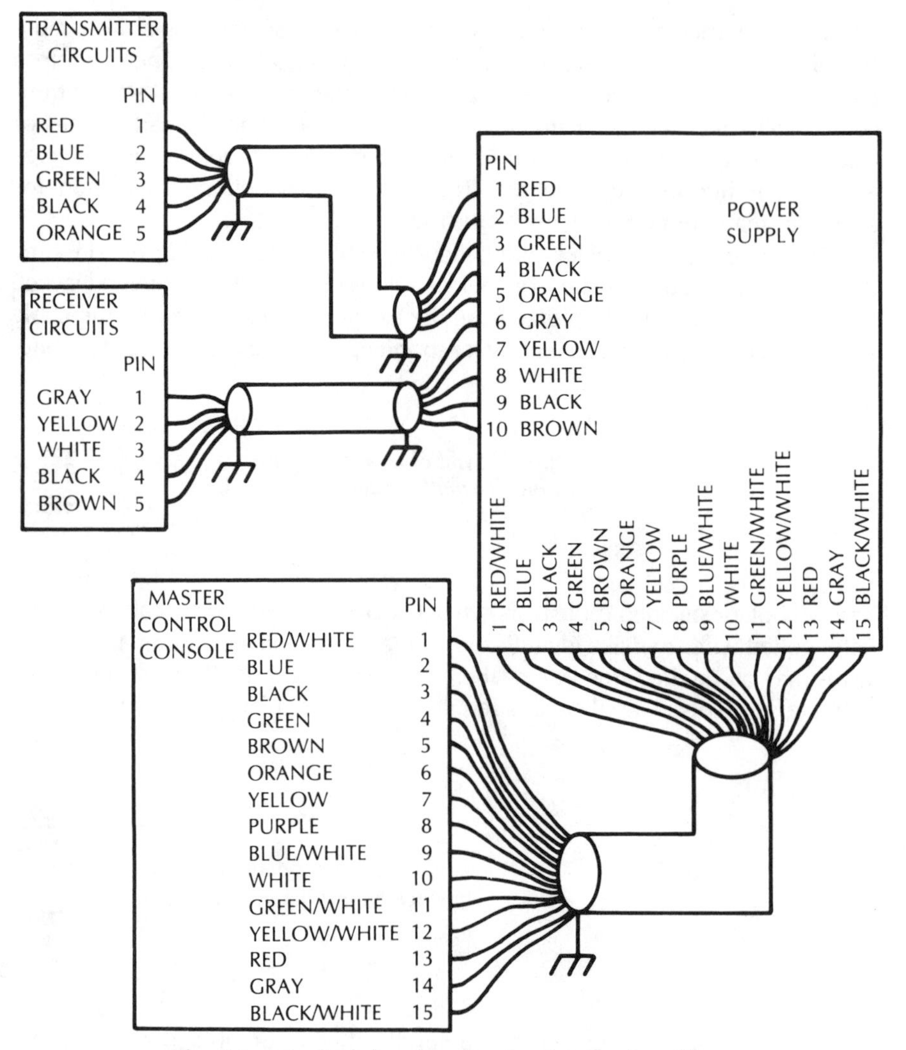

Fig. 5-6. A typical drawing showing inter-unit cabling.

The schematic symbol for a multiple-wire connector is shown in Fig. 5-7. Although the diagram illustrates 6- and 10-pin connectors, the same symbol is used for connectors with more or fewer contacts. Figure 5-8 shows additional types of connectors.

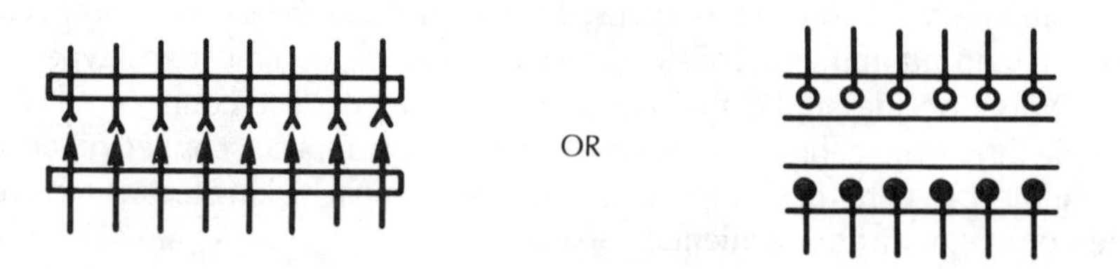

Fig. 5-7. This type of symbol is used to represent multiple-wire connectors. The number of terminals represents the number of individual conductors in the cable.

Fig. 5-8. Notice the multiple-wire connector ports (courtesy of Hewlett Packard).

AUDIO CONNECTORS

An audio-type connector is used to couple audio signals from a microphone, turntable, tape recorder, receiver, or musical instrument to an amplifier. Audio connectors are designed to transfer the audio signals with as little loss or distortion as possible. The cable used to carry audio signals is always a shielded type and may contain one, two, three, or more wires in addition to the shield. In some cases, each lead has its own shield.

Typical audio connectors are shown in Figs. 5-9 and 5-10. The single-contact type uses one pin for the audio and the case for the shield or ground.

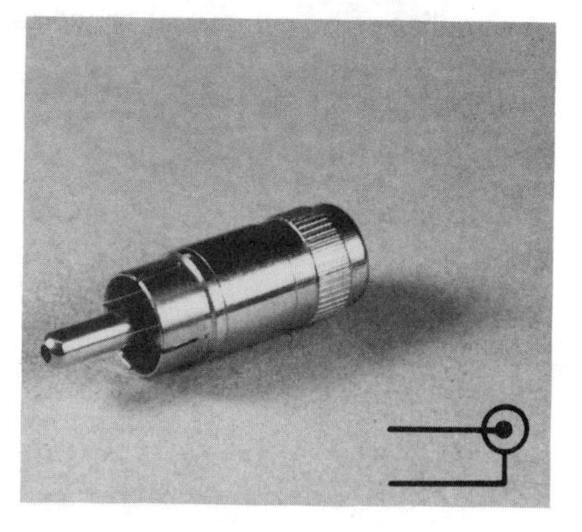

Fig. 5-9. A small solder-type audio connector and symbol.

Single-wire shielded audio lead is commonly used on musical equipment, turntables, tape recorders, hi-fi systems, and on some microphones. A microphone with an on-off switch could use a three- or four-wire audio lead. Multiple-microphone installations and more complex systems used in broadcast studios could require up to six leads in one cable.

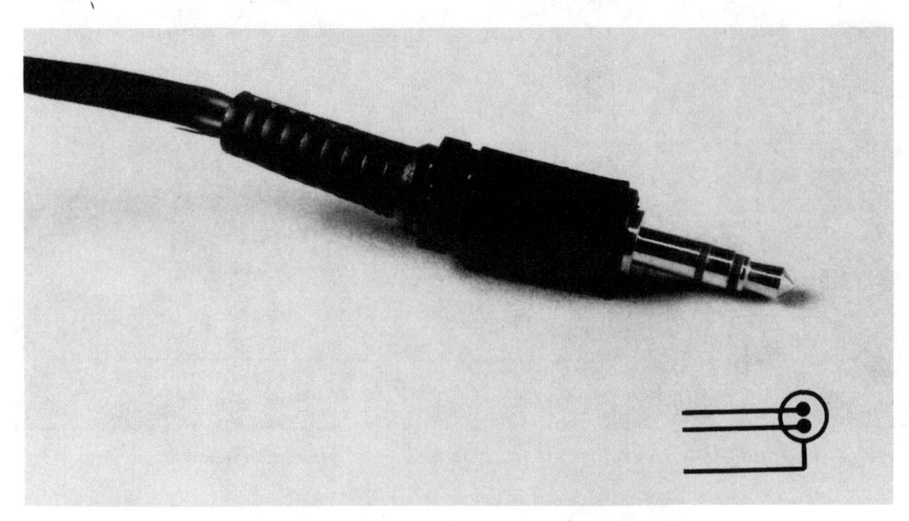

Fig. 5-10. A stereo audio connector and symbol.

SPECIAL CONNECTORS

In addition to multiple-contact and audio-type connectors, there are hundreds of types designed for special applications—connectors to transfer rf (radio frequency) power in vhf/uhf television or in two-way communications, for example. The cable used with these connectors is normally a shielded type called *coaxial* cable. This cable is specially made for various applications, depending on the equipment requirements for power, frequency, impedance, and the length of cable needed. Schematically, coaxial cable looks like any other shielded lead. A sampling of typical coaxial connectors is shown in Figs. 5-11 and 5-12.

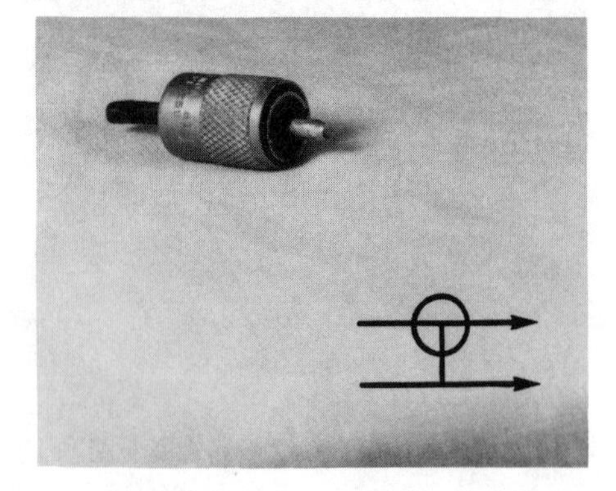

Fig. 5-11. You must solder the connections in this type of rf connector; note symbol.

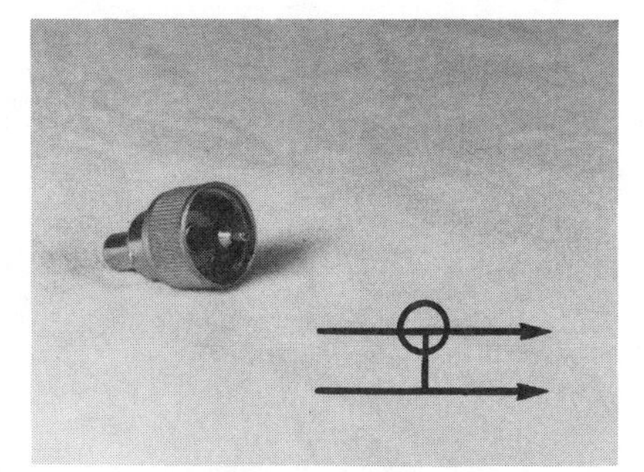

Fig. 5-12. A solderless type of rf connector with symbol.

As with other connectors, there are types for cable-to-cable and cable-to-chassis applications. Coaxial connectors are also designed to transfer the signal with as little loss as possible and are usually made to accept specific sizes of coaxial cable. However, in rf applications, avoiding signal losses is even more crucial than in audio uses because of the frequencies involved. Radio frequency signal levels can be reduced drastically by long cable runs or by improper matching between the cable and equipment. For this reason, both connectors and cable must have low-loss characteristics.

SPECIAL-PURPOSE CABLE

Multiple-wire cable can have any number of individual leads and may or may not be shielded. Figures 5-13 and 5-14 illustrate various types of cable. Most cable has an outer jacket (insulation) of rubber or plastic and the individual wires can be encased in an insulating material. Furthermore, the individual leads in a multi-conductor cable are available in either solid or stranded wire. All variations depend on the application. Solid leads are normally used in equipment for inter-circuit connections, such as telephone cable. If vibration is a factor, as in aircraft wiring, stranded leads are a must because they help prevent breakage.

Fig. 5-13. Multiconductor ribbon cable.

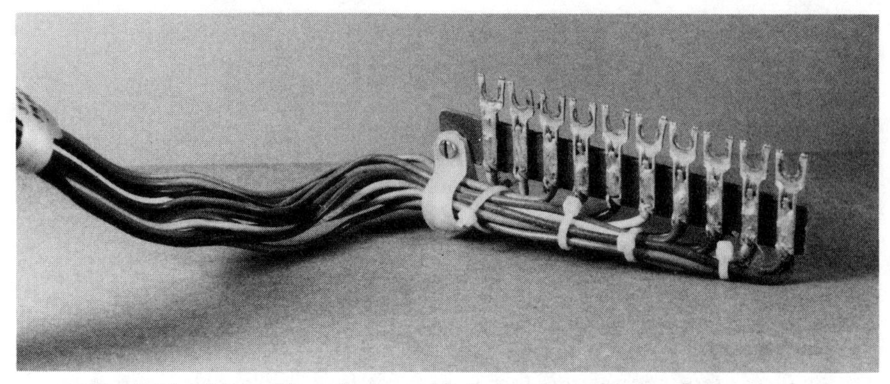

Fig. 5-14. This multiconductor cable is terminated in spade terminals.

Because audio cable and rf cable, as indicated earlier, must be designed to introduce as little distortion and loss as possible, the connectors used must be at least as efficient as the cable. Audio cable usually has one or more conductors, each surrounded by a special insulating material which, in turn, is covered with a metal jacket or shield. The entire assembly is then further encased in a rubber or plastic jacket. Figure 5-15 shows the construction of audio cable.

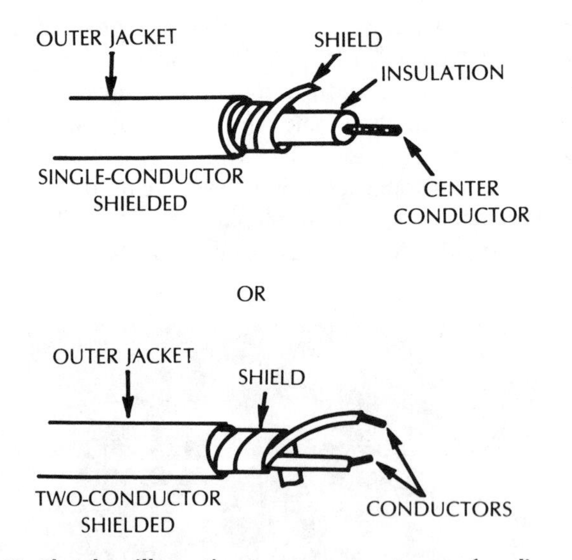

Fig. 5-15. Sketches illustrating two common types of audio cable.

The wire in audio cable is almost always stranded. Coaxial cable is available with either solid or stranded leads. Like audio cable, "coax" consists of a specially insulated center conductor covered with a metal shield and, finally, by a rubber or plastic jacket. However, unlike audio cable, coaxial cable has only one center conductor and a shield as shown in Fig. 5-16. The choice of using either solid or stranded coaxial cable depends on the application.

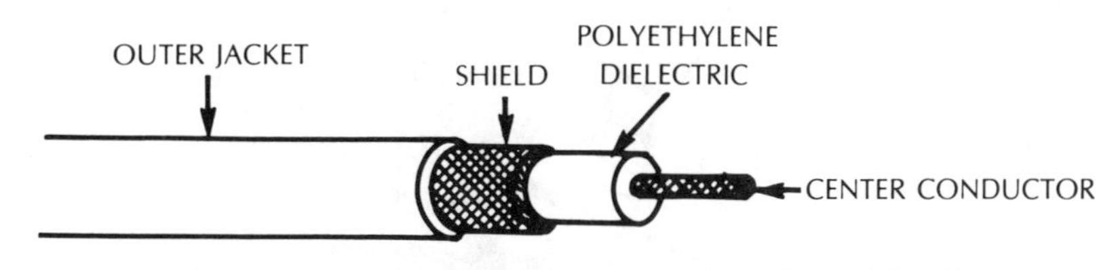

Fig. 5-16. Drawing showing construction of coaxial cable.

Another type of special cable is *twin-lead*—the flat, two-conductor wire used for TV antenna lead-in. It is simply two wires covered with an insulating material that also separates the leads by a specific measure. Twin-lead comes in several types; some are covered by a foil, which acts as shielding. See Fig. 5-17.

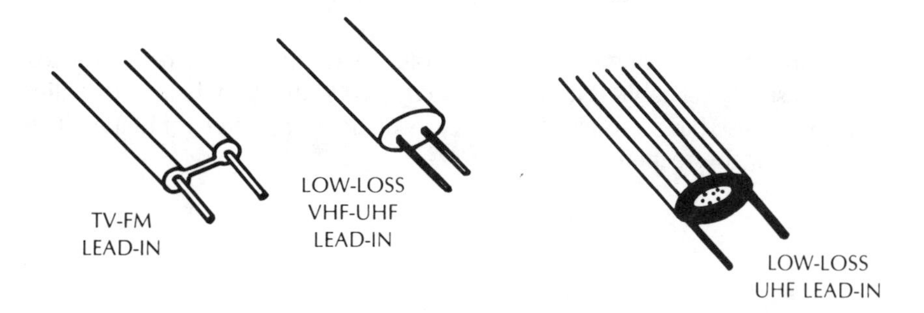

Fig. 5-17. As these drawings show, TV lead or twin-lead is available in several basic forms.

Speaker wire is also a two-conductor cable somewhat like an appliance cord, only it is not made to handle the current required by an appliance. Typical speaker cable is sketched in Fig. 5-18. Usually, each conductor is stranded and insulated with a rubber or plastic jacket. Speaker wire leads are also color-coded (usually one copper and the other silver) to ensure the proper polarity on speakers.

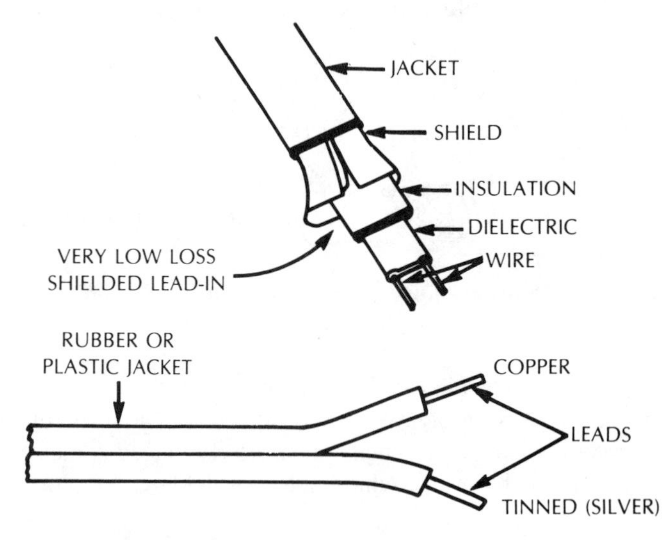

Fig. 5-18. Two types of typical speaker connecting wire.

Finally, there are single-conductor cables made for test instrument leads. In most cases, a test cable is a single lead covered with a very heavy insulating jacket. Some also have a shield. The heavy insulation is required for high-voltage protection for the technician.

6
Types of Diagrams

Before you pack your family into the old bus for a long trip, one of the first things you do is get out the road maps to find the best route to your destination. From experience, we know that road maps use codes or symbols to point out campsites, parks, mileage check points, etc. They might look complicated to a beginner, but after using them a couple of times road maps are really very simple, thanks to a system of symbols. With today's complex highway system, the only way to get across unfamiliar country without getting lost is to use a map.

A home builder uses a road map, too. But a builder's map tells him where to place the frame on the foundation and where to locate supports. This type of map is called a blueprint. An electronic or electrical diagram does the same job for a technician who has to find his way through the complicated maze of wires and parts in a television or radio chassis. Without these valuable diagrams, a technician would spend hours tracing wires from point to point.

SCHEMATIC DIAGRAMS

Several types of diagrams are used in electronics, but the *schematic* diagram is the most widely known because it is used by technicians for service work.

A schematic diagram of a power supply is shown in Fig. 6-1. At first glance, the uninitiated might see nothing but a jumbled mass of lines with a lot of meaningless numbers and letters. That California road map looked like a nightmare at first, too. But a well-drawn schematic, coupled with the right approach, can be interpreted with little difficulty. Usually, the schematic diagram is drawn so the technician follows the signal path from left to right. In other words, the input is from the left and the output is to the right. Components are also numbered in order from left to right. As previously discussed, similar parts have the same letter designation. Symbols make the schematic easier to read, as they do on a roadmap.

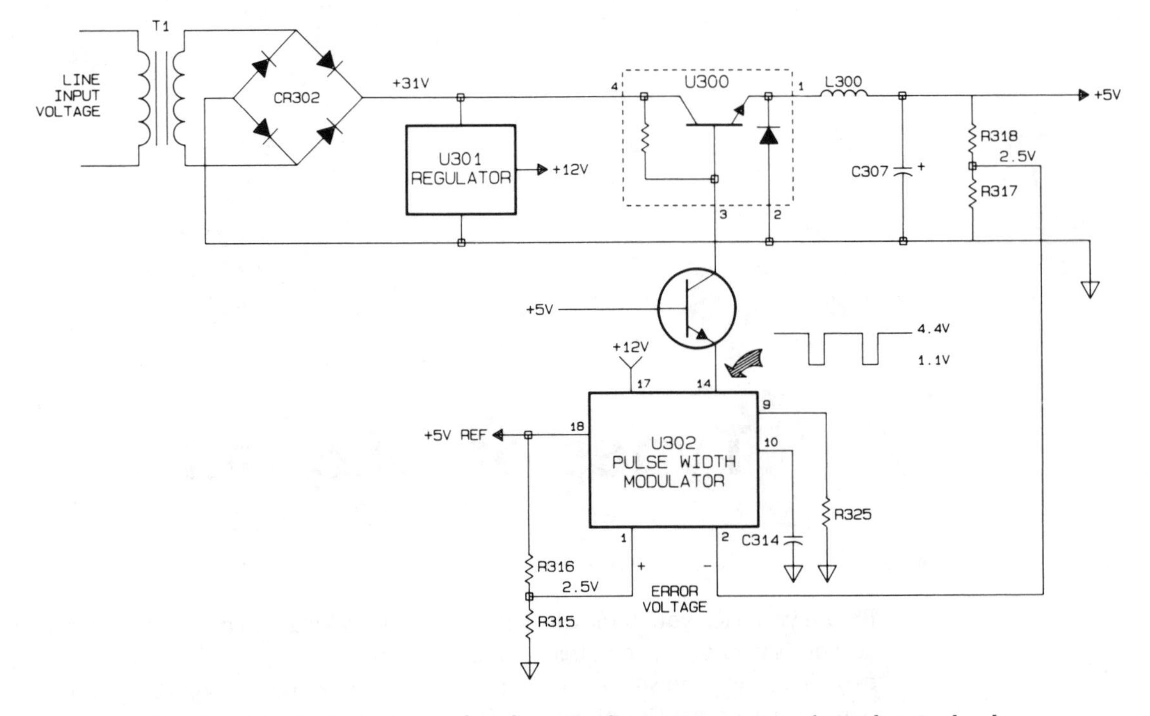

Fig. 6-1. A typical power supply schematic diagram (courtesy of Hewlett Packard).

To identify each component, the manufacturer provides a separate replacement parts list in the service manual that gives a full description of each component. A portion of a typical parts list is shown in Fig. 6-2.

R1 330 OHM
R2 1 KILOHM
C1 .005 MICROFARAD
C2 .01 MICROFARAD
Q1 2N2222
D1 1N4001

Fig. 6-2. A typical parts list.

It should be noted that a service manual is not normally included with the purchase of the equipment. It must be requested, with the schematic, from the manufacturer. Some companies compile schematics and servicing information that they make available through local electronic dealers or on a subscription basis.

In addition to providing a signal-tracing path through a piece of electronic or electrical equipment, a schematic also tells the technician the correct voltage and resistance values at specific points in the circuit. Armed with this information and suitable test equipment, the technician can quickly trace a problem in a TV or radio to the defective component.

BLOCK DIAGRAMS

A typical block diagram is shown in Fig. 6-3. A block diagram is nothing more than a simplified layout of the electronic circuit using squares or rectangles to illustrate the various stages and to show the path of the signal.

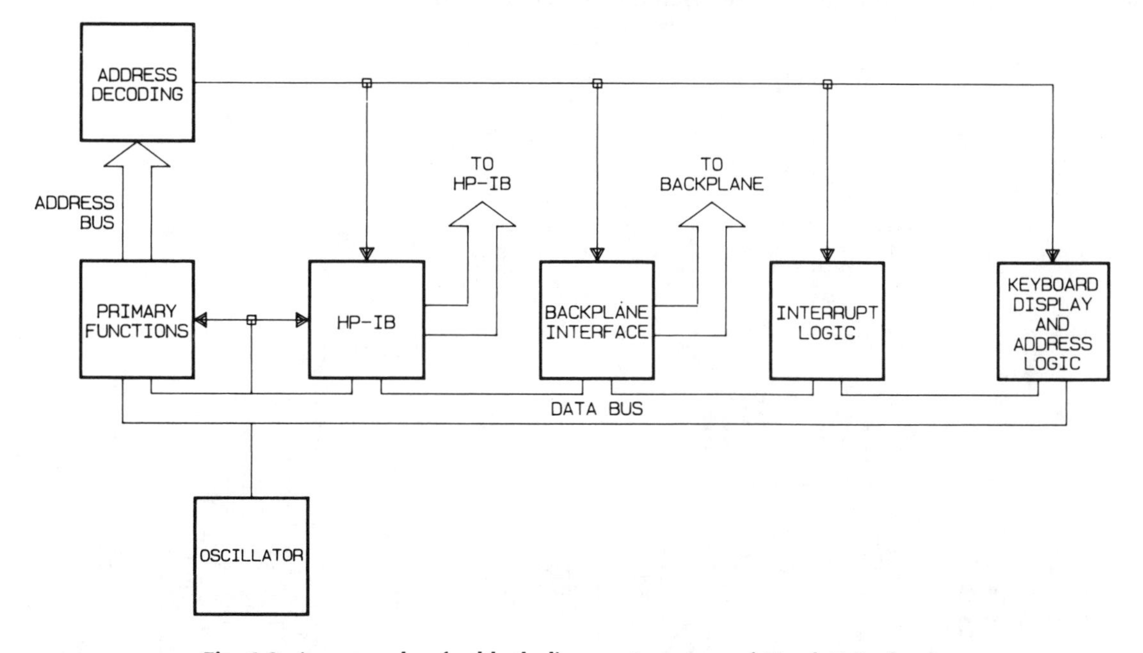

Fig. 6-3. An example of a block diagram (courtesy of Hewlett Packard).

Block diagrams are useful for a quick and easy understanding of the relationship between circuits without the confusion of tracing wires. The various blocks are often labeled to show the tubes or transistors that function in that particular stage and are very helpful in more complex circuits in locating stages that perform several functions. A block diagram is also very helpful in explaining circuit theory and is widely used for this purpose in many basic electronics courses.

LAYOUT DIAGRAMS

A layout diagram shows the physical location of the circuit components. As shown in Fig. 6-4, the diagram indicates the position of ICs, transistors, and connectors. This type of diagram is very helpful to a service technician for locating various test points and adjustments during alignment and parts replacement. It is sometimes called a component placement diagram. Controls or other components shown in dashed lines are located beneath the chassis or below the view being illustrated. Tube diagrams may also have a line or space to show pin orientation to make it easier to re-insert a tube into a socket you can't see.

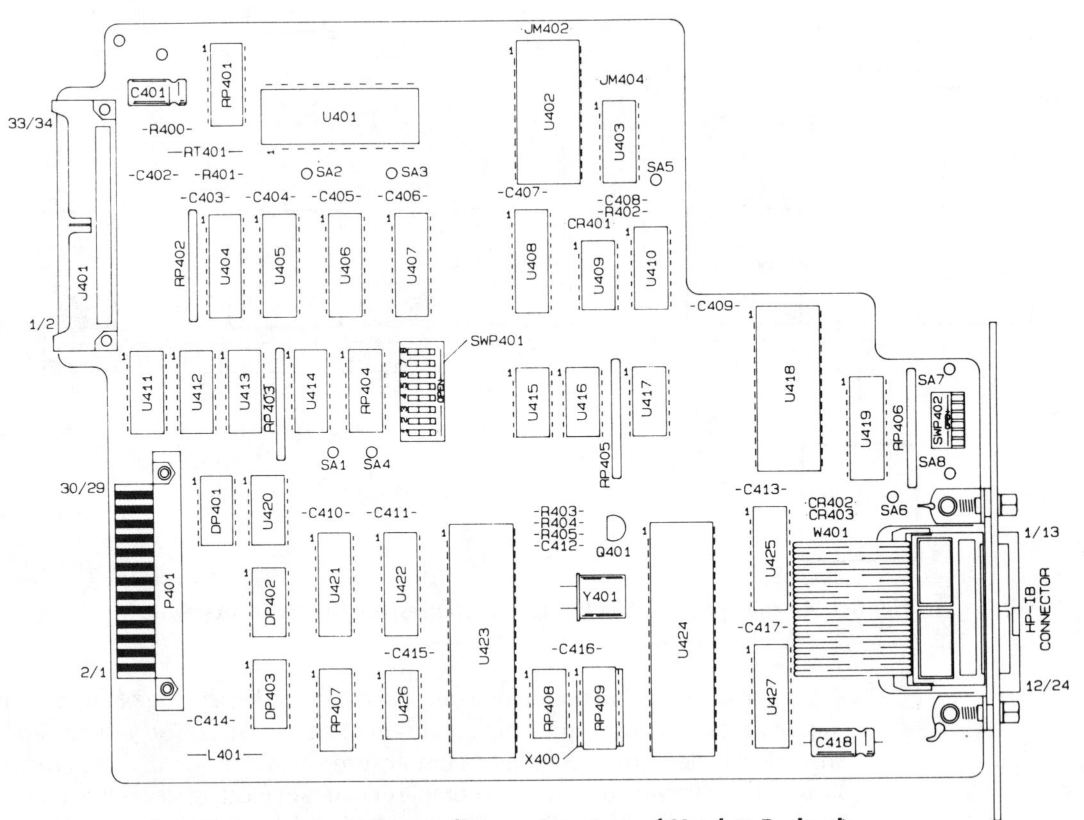

Fig. 6-4. A component layout diagram (courtesy of Hewlett Packard).

Although a layout diagram generally illustrates only the major components in tube equipment, layout diagrams of transistorized equipment might be actual photos or drawings of complete circuit boards. Figure 6-5 shows an IC circuit board.

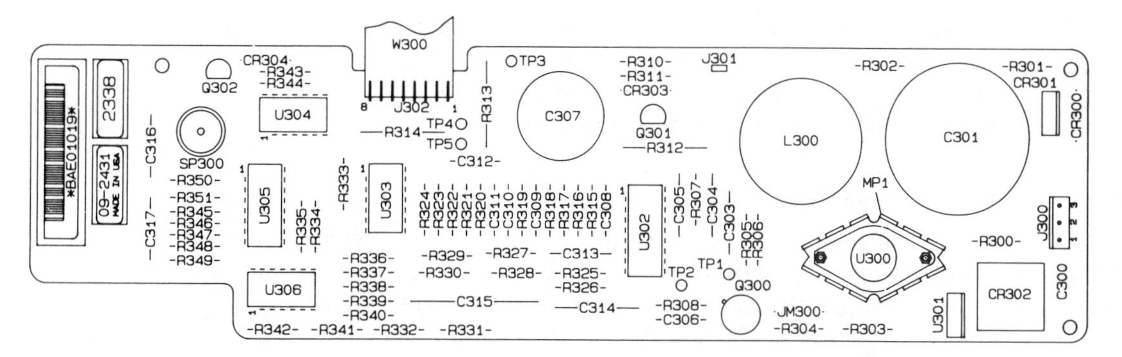

Fig. 6-5. A layout diagram showing the position of ICs, resistors, capacitors and other components (courtesy of Hewlett Packard).

PICTORIAL DIAGRAM

A typical pictorial diagram or drawing is shown in Fig. 6-6. Similar diagrams are often used to aid in building kit-type electronic equipment. The pictorial diagram is suited for this function because it gives the builder a physical view of the actual wiring that is easier for a novice to follow than a complicated wiring diagram. In a rather simple piece of equipment, the pictorial diagram will show all of the components. In more complicated equipment, the components will be installed in groups with a pictorial for each group. This eliminates confusion for both the builder and the manufacturer.

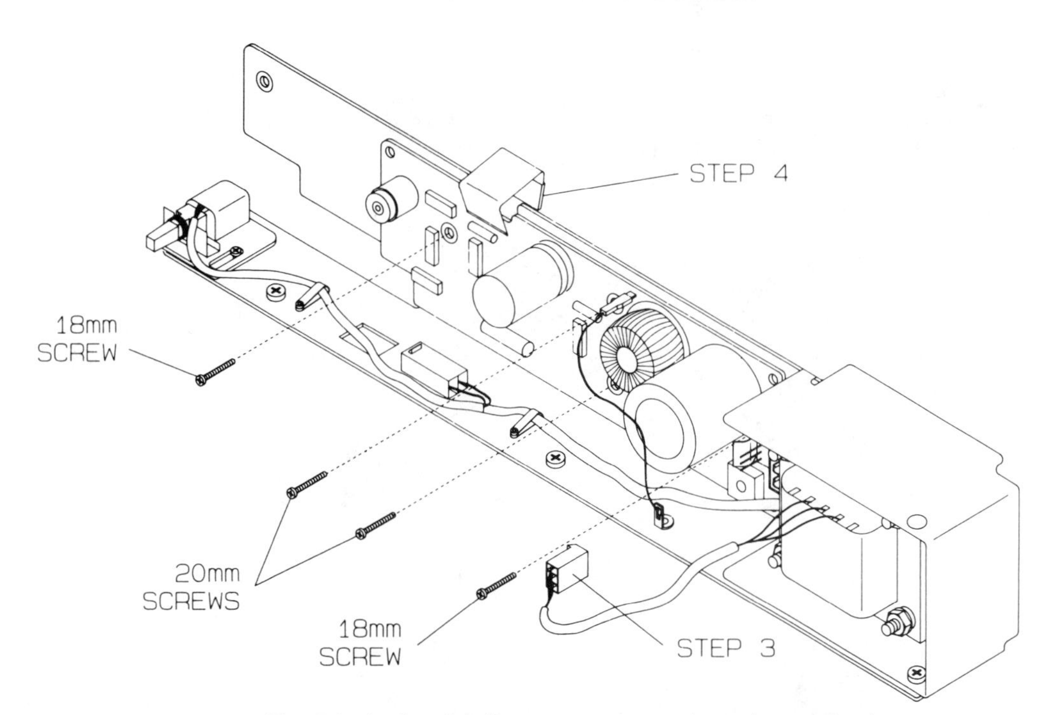

Fig. 6-6. A pictorial diagram (courtesy of Hewlett Packard).

Transistorized and integrated-circuit equipment uses printed-circuit boards to make kit assembly an easy task for almost anyone. The pictorials for these circuits are usually actual circuit board photos or drawings which show all of the components for that circuit, similar to the diagram in Fig. 6-5.

MECHANICAL CONSTRUCTION DIAGRAMS

This type of diagram is used to show the construction of mechanical components and systems which normally cannot be shown on a schematic or other diagram. Mechanical construction diagrams are used to illustrate such things as tape recorder drive systems, antenna rotators, hydraulic systems, etc. An example of a mechanical construction diagram is shown in Fig. 6-7.

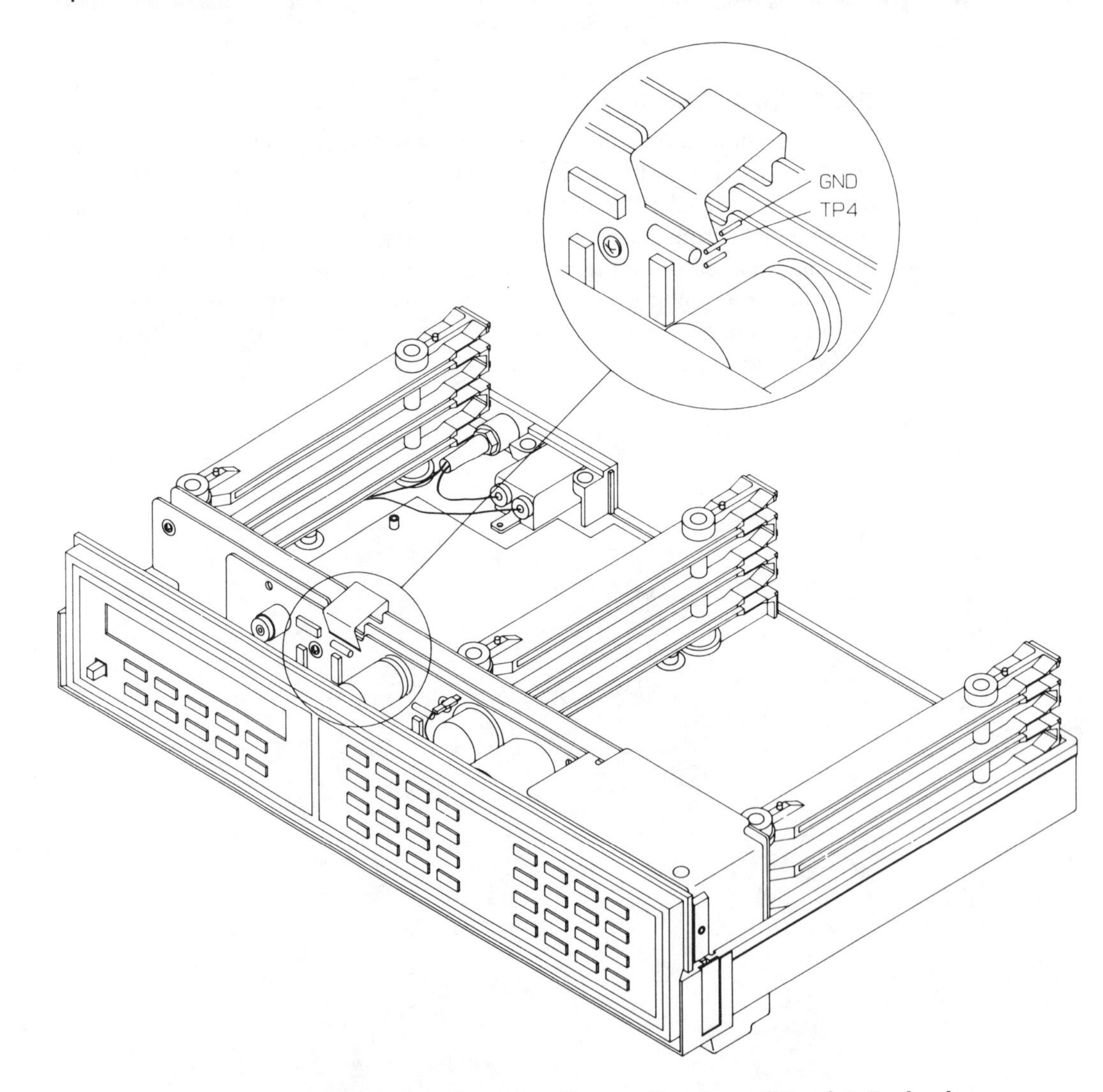

Fig. 6-7. A mechanical construction diagram (courtesy of Hewlett Packard).

Such diagrams are often a drawing, cutaway view, or photo. An exploded diagram showing the arrangement of the circuit board and cabinet is shown in Fig. 6-8.

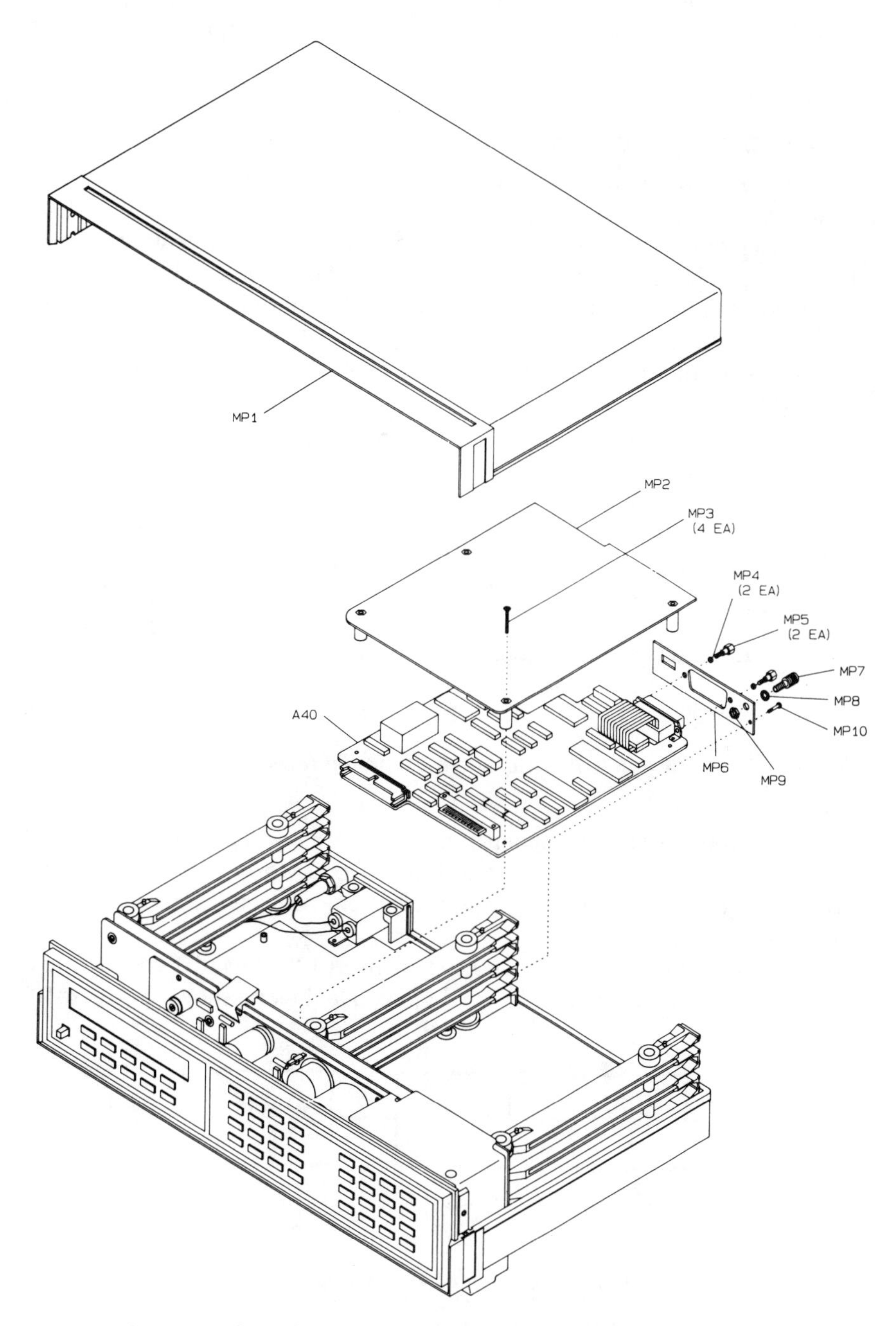

Fig. 6-8. This type of mechanical construction diagram is called an exploded view (courtesy of Hewlett Packard).

DRAWING ELECTRONIC DIAGRAMS

Drawing schematic diagrams is one of the best ways to learn about electronics. The process is not difficult if you follow a few simple rules. This section will show you how to draw both circuit block diagrams and circuit schematic diagrams.

Drawing Block Diagrams

Block diagrams are simple and therefore easy to draw. A block diagram shows the basic stages within a piece of electronic equipment. Figure 6-9 shows a block diagram of a radio receiver. As you can see, no individual components are shown within the separate boxes between the antenna and the speaker. A block diagram like this one can be used to identify the relationship of one stage in a receiver to all of the other stages. If you are designing an electronic device, you should first make such a block diagram. You would then design the actual circuits within each stage.

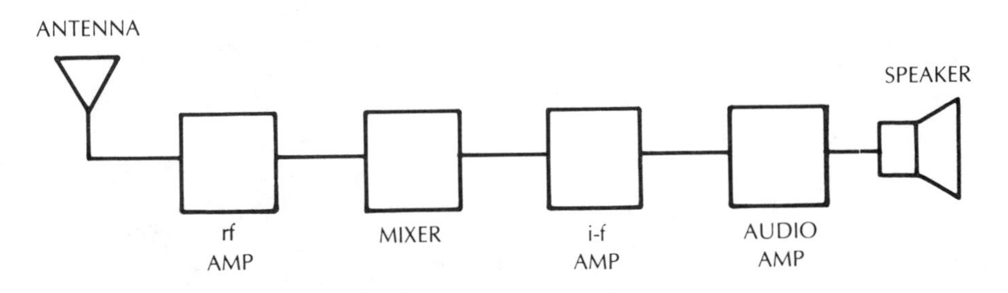

Fig. 6-9. Block diagram of a radio receiver.

To draw your own block diagrams, you must begin by identifying the different stages within the device you intend to build. Don't worry if you don't know all of the stages at first. The design of an electronic device is a progression of design refinements.

Say you want to build a power supply to convert 120 Vac to 12 Vdc. A simple block diagram showing this design would look like the one shown in Fig. 6-10.

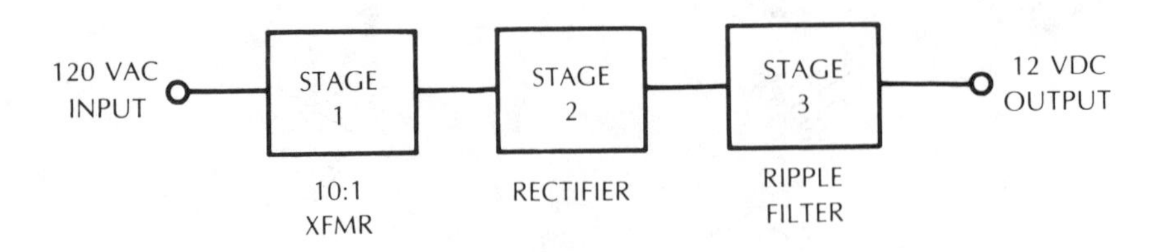

Fig. 6-10. Block diagram of a 12-Vdc power supply.

As the design of your power supply progresses, you might want to change the circuit somewhat. For example, you might decide that you need a very stable output voltage. In this case, you would want to add a voltage regulator. The block diagram would then look like the one shown in Fig. 6-11.

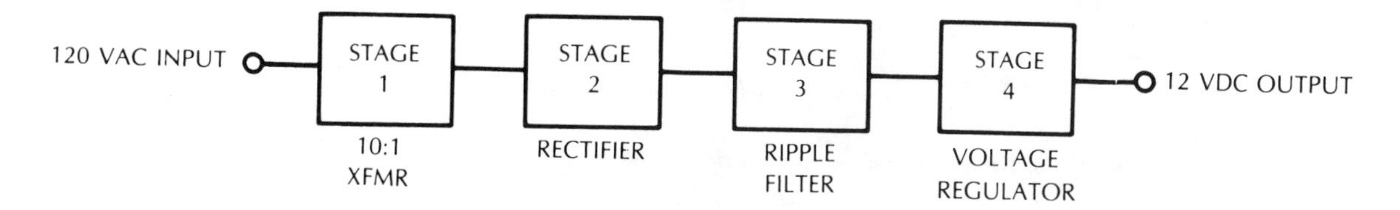

Fig. 6-11. Modified block diagram of the 12-Vdc power supply.

It is very important to make accurate documentation of all phases of your design. Block diagrams help you to visualize the separate stages in an electronic device, and they show the relationship of one stage to another. Should you ever need to troubleshoot the circuit, your documentation will help to refresh your memory concerning the design of the device.

One of the most common mistakes that novice engineers, technicians, and hobbyists make is to jump right into building a new circuit without adequate design planning and documentation. They usually wind up frustrated when the device doesn't work properly, and then a few days later can't remember all the steps they took. This forces them to start all over again from scratch. To avoid this frustration follow the rules below:

- Plan the circuit and make a block diagram
- Make notes on component types and values.
- Save all of your documentation.

(A spiral notebook is best because there are no loose pages to get lost). If you follow the above procedure, you will be able to use your block diagrams, notes, and other documentation to begin the final stage of circuit design. Then you can draw the actual electronic schematic diagram.

Drawing Circuit Diagrams

Schematic diagrams are considerably more complex than block diagrams. These diagrams show all of the components and interconnecting wiring. If you follow a few simple rules, you will have little difficulty with these drawings.

Figure 6-12 shows the schematic diagram of a dc power supply. You will notice that instead of labeled blocks, all of the actual components are shown.

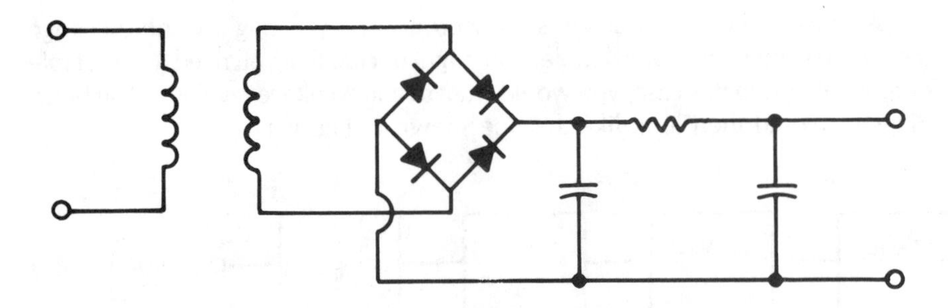

Fig. 6-12. Basic ac-to-dc power supply schematic.

Figure 6-13 shows how the block diagram (Fig. 6-10) is related to the schematic diagram. If you leave out any stages of your block diagram, you will certainly discover them when you try to draw the schematic diagram!

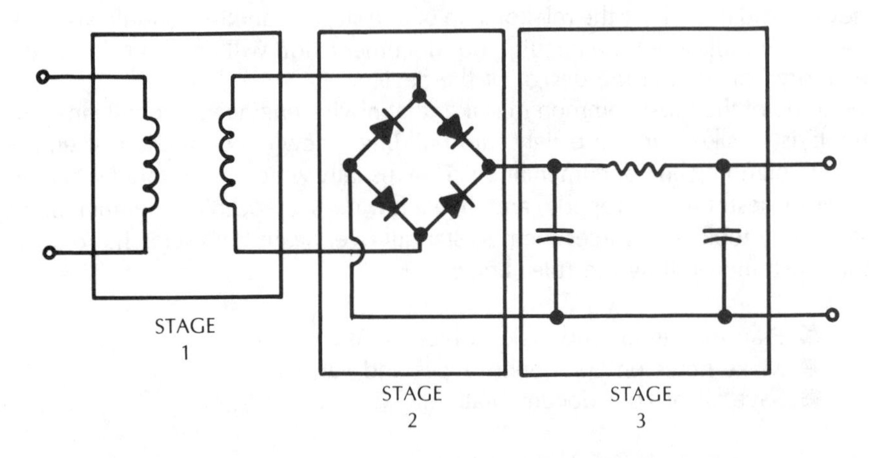

Fig. 6-13. How the block diagram is related to the schematic diagram.

To draw a circuit schematic diagram like the one shown in Fig. 6-13, you must first select the appropriate components. You can use either an existing schematic for a similar device, or you can breadboard the circuit on an experimenter's board. You should make additional schematic drawings as your design changes and is perfected. It's best not to make too many changes to a single drawing because this can cause confusion. Circuit changes can, however, be made using different colors. The original circuit should be drawn in black. Primary changes should be in red and secondary changes in blue. If you need to make tertiary changes, it's time for a new drawing! You usually don't need to redraw the entire circuit, just the stage or portion of the stage that has been changed. However, don't forget to identify it.

The following are some basic rules about drawing schematic diagrams. Like all rules, there are exceptions and not everyone will agree 100 percent with all of them. However, following these rules will help you to be more consistent in drawing schematic diagrams.

Rule 1. Use standard schematic symbols. Over the years, schematic symbols have gradually changed. For example, many older schematic diagrams show coil and transformer symbols, as in Fig. 6-14(A). The modern equivalent symbols are shown in Fig. 6-14(B).

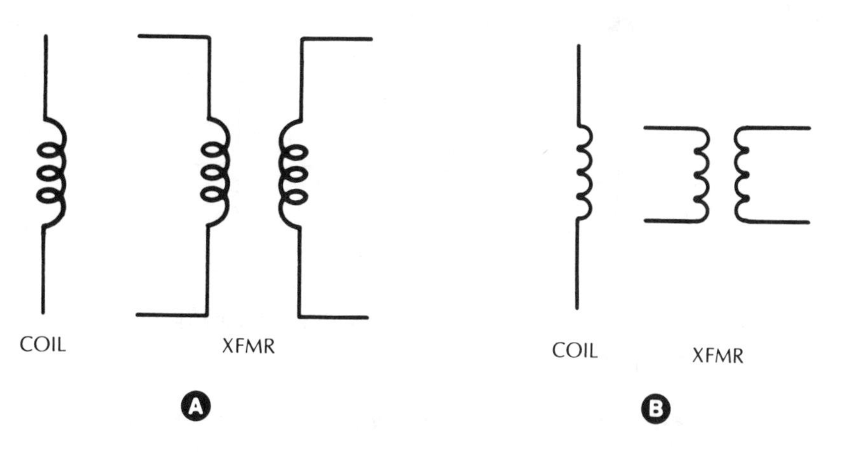

Fig. 6-14. A) Obsolete coil and B) modern symbols.

Use the modern system of units to represent all quantities. The modern system is the *International System of Units.* It is also called the *SI* system (for Systeme International d'Unites).

Rule 2. Be consistent when drawing the interconnecting wiring. Few things in this world are more irritating than carefully following a schematic diagram only to discover that one or more of the wires were shown incorrectly.

There are two different conventions for showing a wiring connection. See Figs. 6-15, A and C. There are also two different ways to show wires that cross each other without any electrical connection. See Figs. 6-15 B and D.

Figure 6-15(A) clearly shows a connection dot at the junction of two wires. Figure 6-15(D) clearly shows that one wire "jumps over" the other. Notice, however, that Figs. 6-15(B) and (C) are identical, although they mean the exact opposite! Why is this so? The reason is that two different conventions are needed for simplicity. In some schematic drawings, it would clutter up the diagram to include connection dots for every single connecting wire. In other drawings, the number of "jumpers" would be excessive.

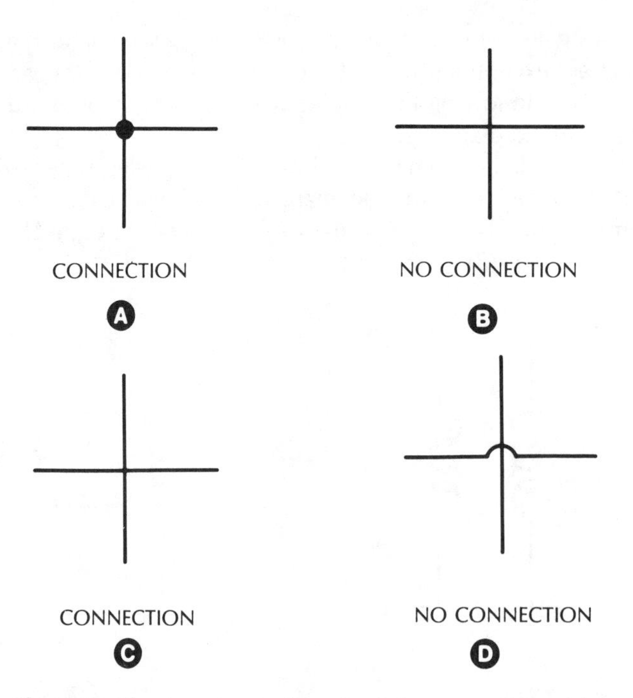

Fig. 6-15. The two conventions for interconnecting wiring.

The two wiring conventions will not cause you any trouble when making your own schematic drawings. This is because you can usually use whichever system suits you—as long as you clearly identify which system you're using, and use it consistently. Some people prefer to use the best of both conventions. Here, you must use a connection dot for *every* connection and a "jumper" for *every* no-connection. It's more work, but there should never be any doubt as to whether there is a connection or not.

Rule 3. Components must be numbered. When numbering components in a schematic diagram, you should always use the standard abbreviations (see Table 6-1). The components should be numbered from *left to right* and from *top to bottom.* See Fig. 6-16. This system of numbering makes it much easier to locate a specific component on a complex schematic diagram.

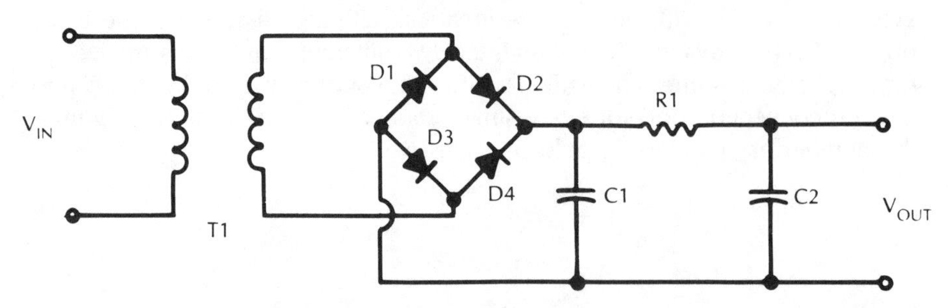

Fig. 6-16. How to number electronic components.

Component	Abbreviation
battery	B
capacitor	C
coil	L
crystal	X
diode	D
integrated circuit	IC
resistor	R
switch	Sw
transformer	T
transistor	Q
vacuum tube	V
zener diode	CR

Table 6-1. Some Standard Component Abbreviations for Use In Schematic Diagrams.

Component abbreviations should always be capitalized and the number should always be on-line with the capital letter. This applies to all "real" components. Theoretical, unknown, or variable quantities such as input voltage, current through a component, or output current always use subscripts. For example, these quantities are indicated as: input voltage (V_{IN}); current through, say a resistor (I_R) and output current (I_{OUT}).

When making any schematic drawing you should always show the inputs, outputs and ground, including all power supply connections. Sometimes the power supply connections are not shown in the manufacturer's literature. It would be redundant to keep showing the same connections over and over again. But for the purposes of drawing your own schematic diagrams, you should show *all* inputs, outputs, and ground connections. Show the value of all components according to the units shown in Table 6-2. If you don't indicate the units on the schematic, you must include a separate *parts list*. You do not need to include the ohm symbol for resistors but you must use the kΩ and MΩ symbols.

Component	Standard Unit	Abbreviation
Resistor	ohm, kilohm, megohm	Ω, kΩ, MΩ
Capacitor	microfarad, picofarad	μF, pF
Coil	millihenry	mH

Table 6-2. Standard Units of Measurement for Use In Schematic Diagrams.

Frequently, the ''μF'' will be omitted for capacitors, but the pF should always be denoted. If you choose to omit any of these abbreviations, you must clearly indicate this on the drawing. A standard notation might read, ''All resistors are ohms and all capacitors are microfarads unless indicated otherwise.''

If you follow the above rules and remember to keep accurate documentation, you will soon be able to draw more and more complex circuit diagrams. Thumbing through your notebook (or notebooks!) you will be amazed at your own progress.

7
Radio and TV Schematics

The portable AM radio is certainly one of the most common pieces of electronic equipment in the world today. The basic design for the circuits in these radios was perfected during the days of vacuum tubes, and the similarities between vacuum-tube circuits and transistorized circuits are very interesting.

THE HEATHKIT PORTABLE AM RADIO, MODEL GR-1009*

This chapter examines a portable AM radio that you can build yourself. It is the eight-transistor Heathkit Model GR-1009. A complete kit containing all the components needed to build this radio can be obtained from The Heath Company, Benton Harbor, Michigan 49022.

While most everyone knows how to turn on a radio and tune in a station, many people never think about how that voice or music is sent through the air to come out the speaker of their radio. However, not everyone has built their own AM radio.

This circuit description is presented in two parts, because you might want to know how it works in general rather than in complex electronic terms. The first part is basic AM radio theory for those who want just the fundamentals. The second part, intended for those with an understanding of electronics, explains in detail how each electronic circuit functions in the radio.

Basic AM Radio Theory

Sound—Audio Signals. Sounds, such as voice, music, bells, car horns, etc., are vibrations that carry through the air as *sound waves.* When these sound waves

---

reach your ear, they vibrate the ear drum and are heard. You can feel sound vibrations if you place your finger on the cone of a loudspeaker or hold a thin piece of paper or metal in front of your mouth as you talk or sing.

A microphone changes sound-wave vibrations into electrical impulses of the same characteristics or wave pattern. These impulses are called an *audio signal* and are amplified for use in various ways. One use is to drive a public address or intercom loudspeaker, which changes the electronic audio signal back into sound waves. The audio signal can also be recorded on plastic disks or magnetic tape for replay later. In the case of radio transmission, amplified audio signals from a microphone, phonograph record, or tape player are used to *modulate,* or impress sound upon, a radio station carrier frequency and sent out over the air.

Audio signals, or those that the human ear can detect, have a frequency range of 20 to 20,000 cycles per second, denoted as hertz (Hz). Ordinary sound waves travel only a few yards, while a loud fog horn might travel a few miles. Finally, electronic, amplified audio signals can be sent over wires (such as your telephone line) as far as the wires will take them.

The Radio Station. A radio station takes the audio signal from a microphone, phonograph record, or recorded tape and sends it out through the air on a radio frequency carrier in the following manner.

Radio frequencies are measured in kilohertz (kHz) or megahertz (MHz), formerly called kilocycles per second and megacycles per second, respectively. *Kilo* means thousand and *mega* means million. The radio frequency spectrum ranges from about 50 kHz to over 1000 MHz, which is much higher than the audio range. Different bands within the radio frequency spectrum are assigned by the Federal Communications Commission (FCC) for various services. For example, the AM broadcast band is from 535 kHz to 1605 kHz, while the citizens band (CB) is around 27 MHz. See Fig. 7-1 for frequency spectrum sample.

Each radio transmitter (station) is licensed by the FCC to operate at a specific frequency in its band and is assigned "call letters" that identify the station. The FCC also regulates the amount of power the station may put out, which has an effect on the distance that the radio frequency (rf) signal will travel.

The radio frequency (rf) is generated at the radio station by an electronic *oscillator*. This rf is fed through a *modulator* that imposes the sound on the rf, and then through an rf amplifier that amplifies it to a usable level. This signal is then routed to an antenna that radiates the rf signal into the surrounding air. The rf signal itself is known as a *carrier* for the audio signals that modulate (or modify) it.

Amplitude modulation (AM) and frequency modulation (FM) are two methods of placing audio signals on a radio frequency carrier. In AM radio, the amplitude (strength) of the rf carrier is modulated (controlled) by the strength or voltage level of the audio signal. In FM radio, the frequency of the rf carrier is made to deviate above and below a center frequency in proportion to the strength of the audio signal.

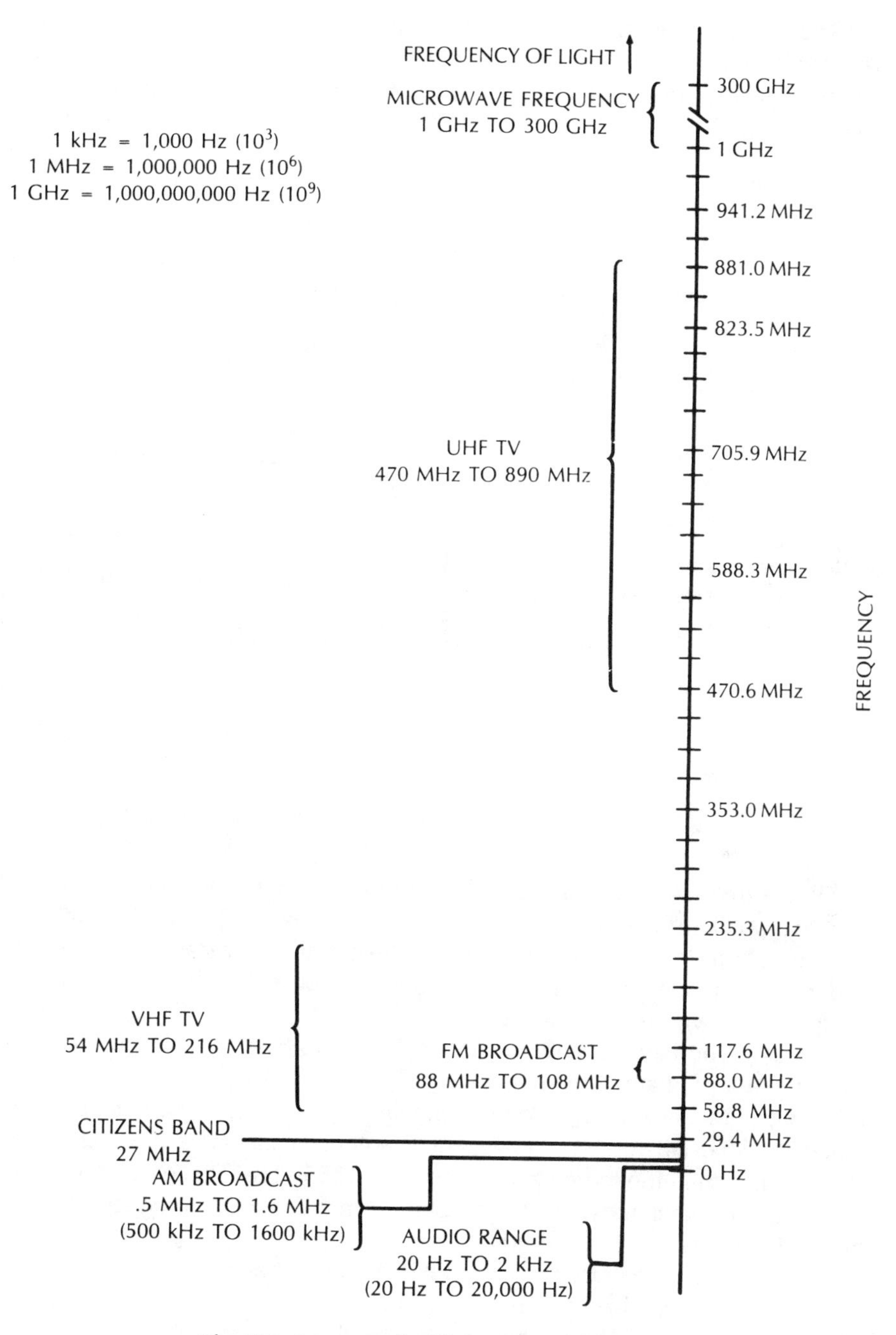

Fig. 7-1. Lower end of frequency spectrum.

The Radio Receiver. This portable AM radio is a *superheterodyne* receiver which has an rf amplifier, an oscillator/mixer, intermediate frequency (i-f) stage, detector, audio amplifier, and audio output stages. See AM Radio Receiver Block Diagram (Fig. 7-2).

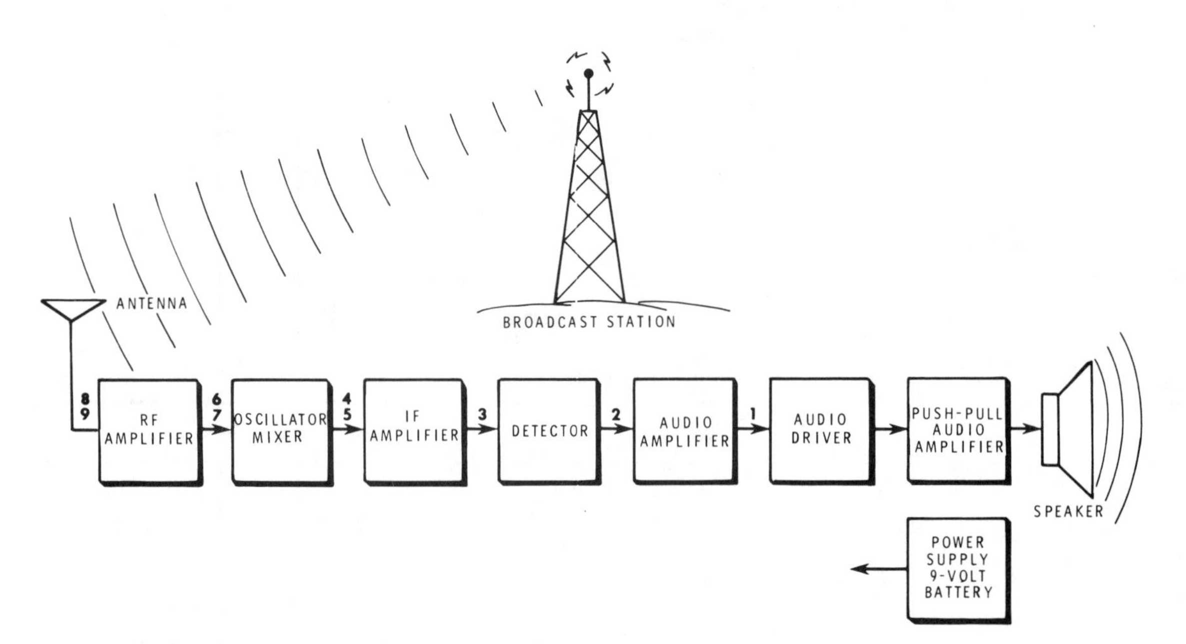

Fig. 7-2. AM radio receiver block diagram.

Radio frequency signals that are radiated from the radio station antenna induce a small current in the antenna of a radio receiver. When you "tune in" a radio station, the circuits in the receiver reject all other station frequencies and amplify the station signal that is tuned in. Following is a description of how a superheterodyne receiver works.

Assume you have tuned your AM radio dial to "14" (1400 kHz) and are listening to a program. The antenna and rf amplifier circuits pass and amplify this rf signal to a mixer stage. At the same time, the setting of the radio dial has tuned its own oscillator circuit to generate another signal that is 455 kHz higher in frequency than the radio station signal, or 1855 kHz. This oscillator signal "beats against" (heterodynes) the input signal in the mixer to produce an intermediate frequency (i-f) of 455 kHz.

All of this "tuning" takes place in the antenna, rf amplifier, and oscillator circuits. The intermediate frequency (i-f) stage is set at 455 kHz and amplifies the i-f signal further before applying it to the detector circuit.

The detector circuit *demodulates* the i-f signal, leaving only the audio signal, which is then amplified in the audio driver stage and in the audio output stage. The loudspeaker then converts the amplified audio signal into sound waves with the same characteristics as those that entered the microphone initially.

Technical Description

Note: This portion of the circuit description is intended for persons with an electronics background. Refer to the AM receiver block diagram (Fig. 7-2) and the schematic diagram (Fig. 7-3) as you read this circuit description.

This portable AM radio is an eight-transistor, superheterodyne receiver that tunes to standard AM (amplitude modulated) broadcast stations from 535 kHz to 1605 kHz. The circuits in this radio include: rf amplifier Q1, oscillator-mixer Q2, i-f amplifiers Q3 and Q4, detectors D2 and D3, audio amplifier Q5, audio driver Q6, and push-pull output amplifier, consisting of Q7 and Q8. Each of these circuits will be described separately. Figures 7-4 through 7-6 show the layout of the radio's components.

Tuner—rf Amplifier. Radio frequency (rf) signals transmitted by broadcast stations induce a small current in the ferrite rod antenna (T1). The desired rf signal is selected by variable tuning capacitor C1 which, in parallel with the primary winding of the antenna coil, tunes the antenna circuit. This selected rf signal is coupled through the secondary winding of T1 to the base of rf amplifier transistor Q1. Bias voltage is applied to transistor Q1 (through resistor R1) and the AGC (automatic gain control) circuit. The amplified signal is coupled from the collector of transistor Q1 through capacitor C4 to the base of oscillator-mixer transistor Q2.

Oscillator/Mixer. The amplified rf signal is mixed with a local oscillator (LO) signal in transistor Q2. Oscillator coil T2, along with capacitors C8, C8A, and C9, determine the frequency of the oscillator signal. Because oscillator tuning capacitor C8 and antenna tuning capacitor C1 rotate on the same tuning shaft, the oscillator frequency is always 455 kHz higher in frequency than the rf signal. Because of the heterodyning of the rf and oscillator signals, the mixer produces signals both higher and lower than the oscillator signal. That is, it produces the sum of the rf signal and local oscillator (rf + LO) and the difference (rf − LO). The difference signal is the i-f of 455 kHz. I-f transformer T3 contains a tuned circuit that passes only the 455 kHz i-f signal to filter F1.

I-f Amplifiers. Filter F1 passes only 455 kHz and rejects, or filters, other frequencies. This filtered i-f signal is applied to the base of i-f amplifier transistor Q3, which is biased by voltage divider resistors R13 and R14. Transistors Q3 and Q4 are direct-coupled and amplify the i-f signal before applying it to a second i-f filter, F2, which further filters the i-f signal.

Detector. Detector diodes D2 and D3 convert the i-f signal into a dc voltage that varies in amplitude at an audio frequency rate. During the conversion, diode D2, resistors R22 and R24, and capacitors C15 and C17 remove the 455 kHz portion of the i-f signal. The audio signal that remains has a dc level that is proportional to the strength of the received radio station signal. This dc voltage is used for automatic gain control as explained in the following paragraph.

SCHEMATIC OF THE HEATHKIT® PORTABLE AM RADIO MODEL GR-1009

NOTES:

1. ALL RESISTORS ARE 1/4 WATT UNLESS MARKED OTHERWISE. RESISTOR VALUES ARE IN OHMS OR K OHMS (K=1000).
2. ALL RESISTORS ARE 5% UNLESS MARKED OTHERWISE.
3. ALL CAPACITOR VALUES ARE IN µF UNLESS MARKED OTHERWISE.
4. REFER TO THE "CIRCUIT BOARD X-RAY VIEW" FOR THE PHYSICAL LOCATION OF PARTS.
5. ▽ THIS SYMBOL INDICATES A CIRCUIT BOARD GROUND.

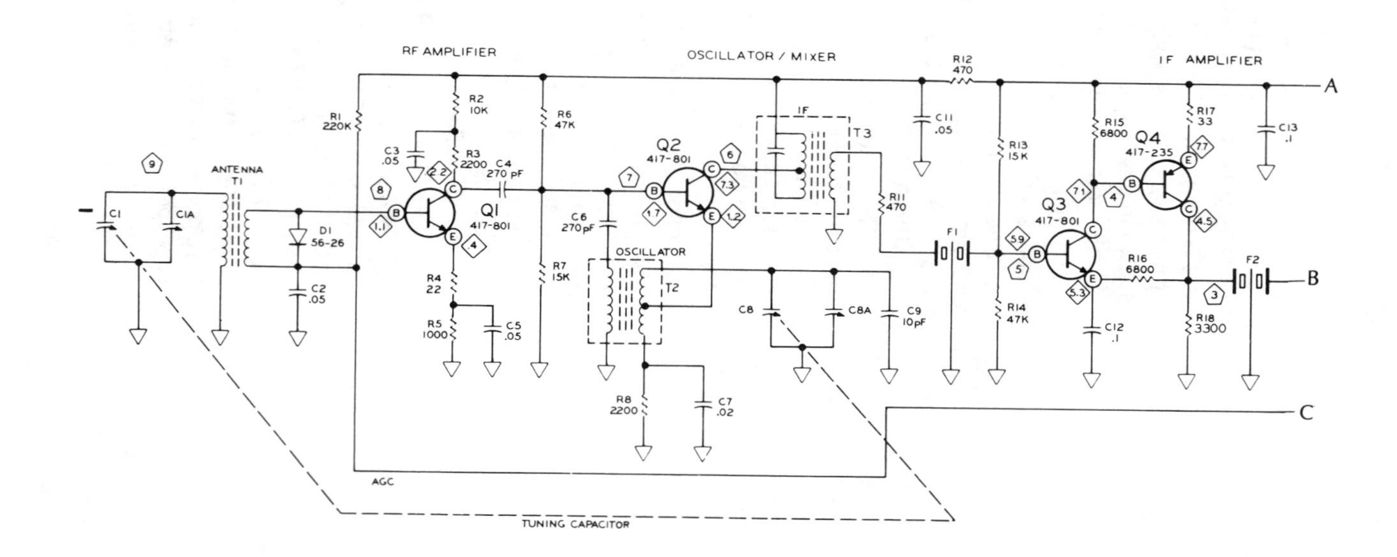

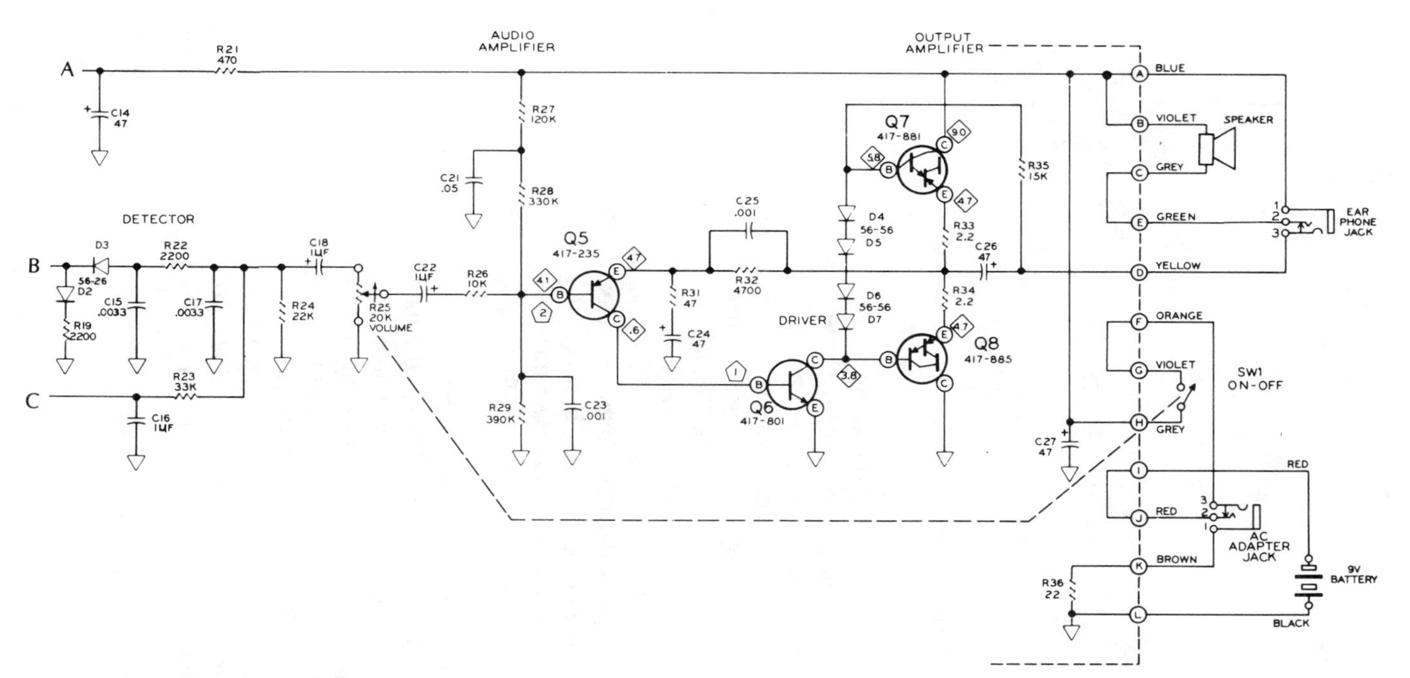

6. THIS SYMBOL INDICATES A TEST POINT. SEE THE POINT-TO-POINT TEST CHART FOR THE FOIL LOCATIONS OF THESE TEST POINTS.

7. THIS SYMBOL INDICATES A POSITIVE DC VOLTAGE MEASUREMENT TAKEN UNDER A NO SIGNAL CONDITION (NO BROADCAST STATION BEING RECEIVED). MEASURED WITH A FRESH 9-VOLT BATTERY INSTALLED. ALL MEASUREMENTS TAKEN WITH AN 11 MEGOHM INPUT IMPEDANCE VOLTMETER FROM THE POINT INDICATED TO CIRCUIT BOARD GROUND. VOLTAGES MAY VARY AS MUCH AS ±20%.

8. THIS SYMBOL INDICATES CLOCKWISE ROTATION OF A CONTROL WHEN VIEWED FROM THE KNOB END.

9. THIS LINE REPRESENTS A MECHANICAL CONNECTION BETWEEN THE POINTS INDICATED.

10. THIS SYMBOL INDICATES A LETTERED CIRCUIT BOARD CONNECTION POINT.

Fig. 7-3. The AM radio schematic.

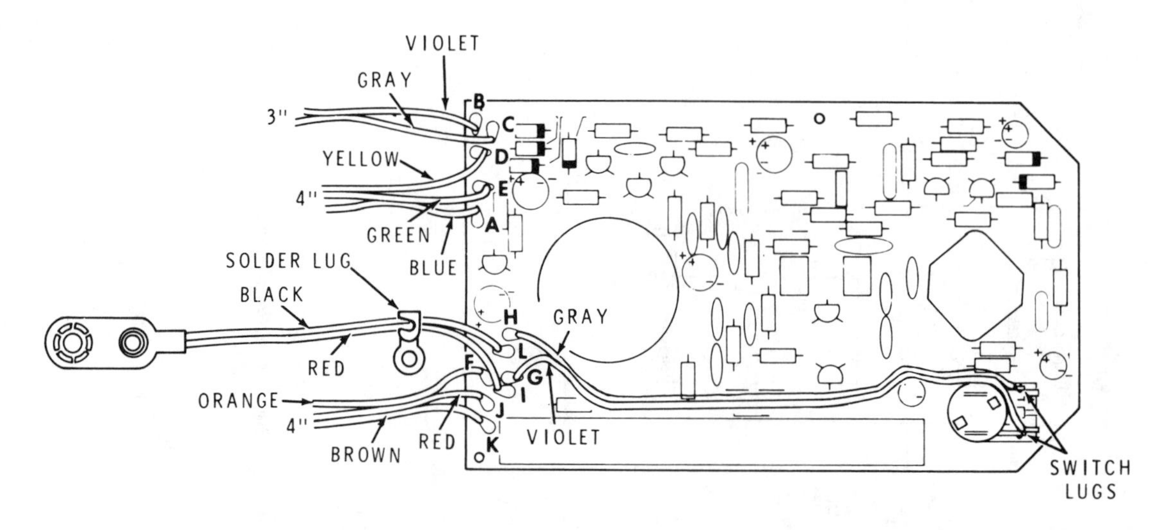

Fig. 7-4. AM radio pictorial showing the color-coding of the wiring.

AGC (Automatic Gain Control). To prevent strong radio signals from being too loud, or weak signals too low to hear, an automatic gain control (AGC) circuit uses the dc component of the detected audio signal to control the gain of the rf amplifier transistor Q1. Resistor R23 "picks off" the dc component from the detector circuit. Capacitor C16 filters the audio and prevents sudden changes in the AGC voltage. This voltage is applied to the base of the rf amplifier transistor Q1 to vary its bias (and gain) inversely with the strength of the received rf signal.

Audio Amplifier—Audio Driver. Capacitor C18 removes the dc voltage from the detected audio signals. The audio signals from the detector are then coupled through volume control R25, capacitor C22, and resistor R26 to the base of audio amplifier transistor Q5, which is direct-coupled to the base of driver transistor Q6. Transistor Q5 amplifies the audio signal and transistor Q6 drives the output amplifier.

Push-Pull Output Amplifier. Driver transistor Q6 is direct-coupled to the base of transistor Q8, and—through diodes D4, D5, D6, and D7—to the base of transistor Q7. This is a push-pull circuit that provides the current amplification needed to drive the loudspeaker. The push-pull action occurs as transistors Q7 and Q8 alternately conduct or are cut off by the negative-going or positive-going portions of the audio signal from the collector of driver transistor Q6.

A negative-going audio signal at the base of transistor Q8 causes the transistor to conduct. When the audio signal at the base of Q8 goes positive, the transistor is cut off and Q7 conducts. Capacitor C26 charges to half the supply voltage when the radio is first turned on. Its high capacitance prevents it from charging or discharging when the speaker is driven at audio frequencies.

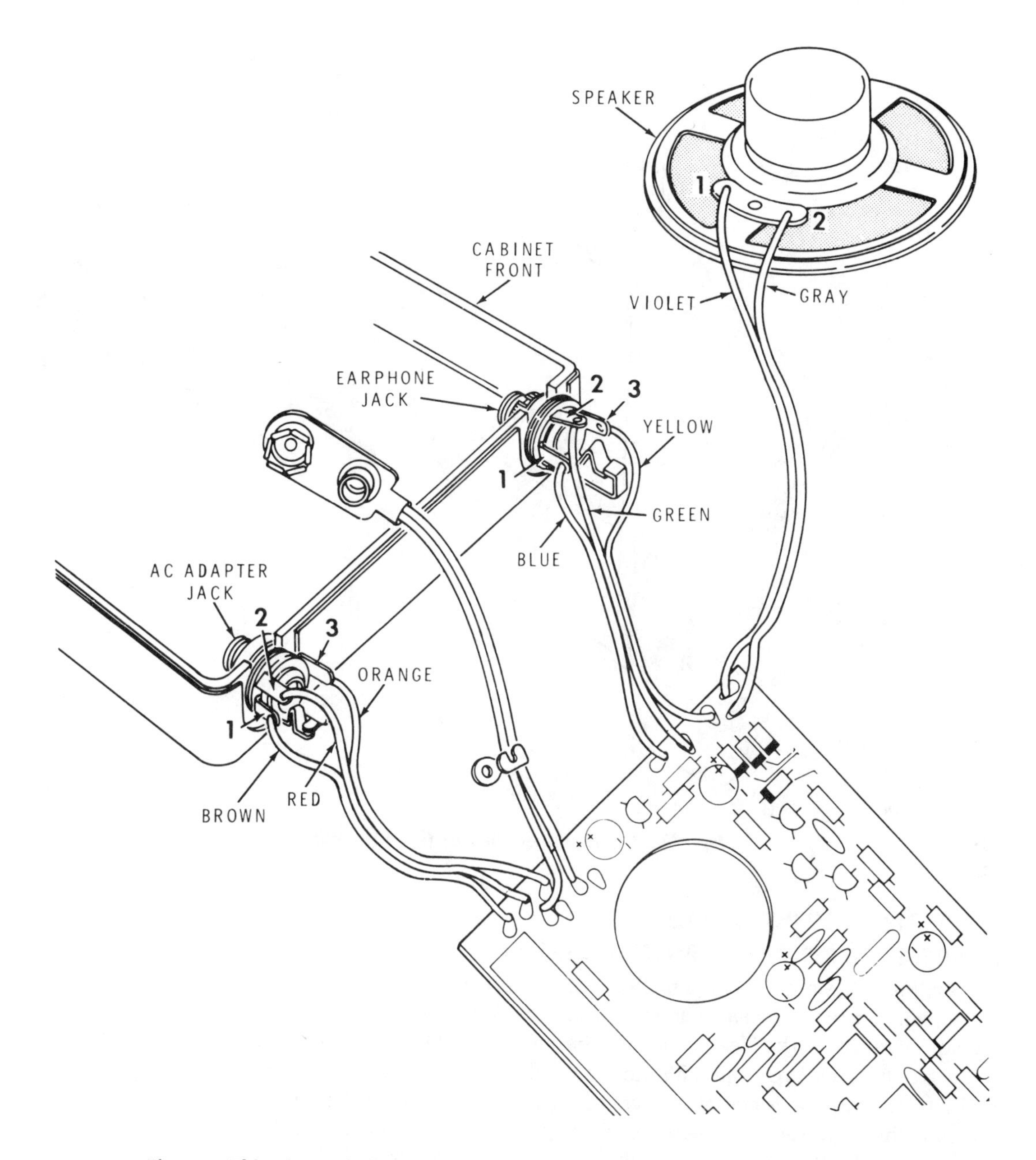

Fig. 7-5. This pictorial shows the wiring of the ac adapter jack and earphone jack.

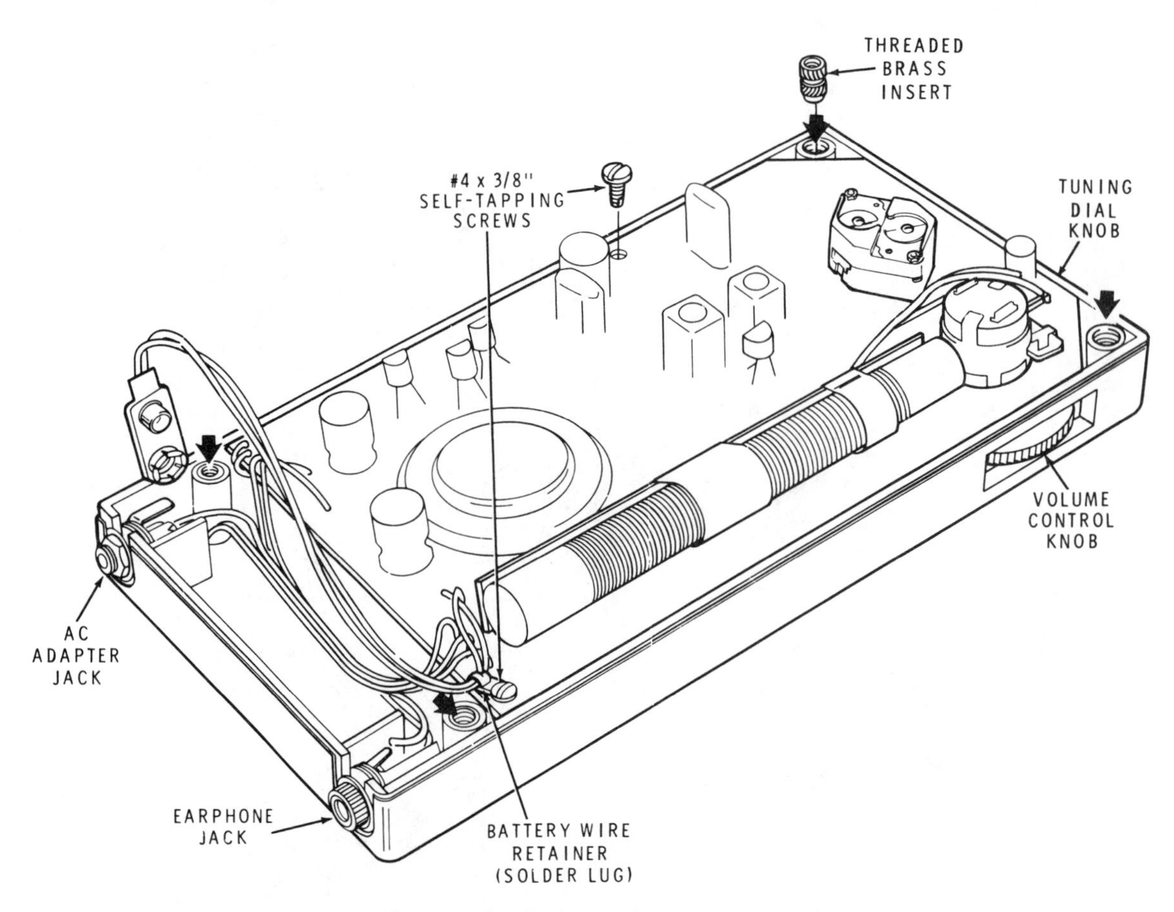

Fig. 7-6. The final assembly of the AM radio.

THE HEATHKIT 19-INCH COLOR TV WITH REMOTE CONTROL, MODEL GR-1903*

By starting with simple schematic diagrams, it is possible to work your way up to the point where you can read and understand a complex schematic diagram for a color TV receiver. The Heath Company has made this process much easier than it was only several years ago. It is now possible to actually build your own color TV from a kit. The schematic diagrams for the TV are broken down into the modules they represent. There are also many pictorials, and even x-ray views are provided in the Heathkit literature that accompanies this color TV kit. As with all Heath products, all of the components and other parts, including the housing, are included. There is no better way to learn about how a color TV receiver works than to actually build one. The material in this section was provided courtesy of the Heath Company.

The Heathkit 19-inch color TV model GR-1903 is really very easy to assemble. Most of the circuit boards and all of the crucial assemblies have been preassembled and are factory aligned. This set will provide excellent performance, easy operation, and give solid-state reliability. Many of the features of this set are Heathkit exclusives. The standard features include the following:

New Computer Brain. In a color television set, the blue, red and green electron guns in the picture tube (CRT) begin to wear out through extended use, producing a washed-out picture with unnatural colors. However, in this set, the new CRT tracking system automatically maintains the colors and keeps them at the original level throughout the life of the picture tube. The colors remain like new year after year.

Quartz-Controlled Electronic Tuning. Electronic tuning ends the need for fine-tuning adjustments. The tuner instantly seeks out and locks onto the exact frequency to which you are tuned. The picture always remains sharp and clear. Because the tuner is electronic, there are no moving parts to wear out and no contacts to corrode.

Timed-Entry Channel Selection. This feature allows you to select a channel, and after a three second delay, the channel will be tuned in automatically. For an immediate selection entry, select the channel and press the ENTER button.

Infrared Remote Control. This allows you to turn the television on and off, change the channel, control the volume level, recall the channel and time display, and operate the optional antenna switch. Infrared operation improves reliability and reduces inadvertant activation due to noise. The remote control also operates Zenith VHS videocassette recorders.

Up/Down Channel Scanning. This feature allows you to scan your favorite channels in an up or down sequence. Once programmed, a microcomputer "remembers" your favorite channels for tuning ease and convenience.

On-Screen Channel Number Display. The display shows you the channel number you are watching and the time of day. The display will show for approximately 4½ seconds whenever you change channels or whenever you press the RECALL button on your remote control unit or the TV's main keyboard.

178-Channel Capability. This allows you to select all VHF/UHF and 122 cable channels, including HRC cable, without the need for an external converter (except for scrambled programs). It selects all 68 standard broadcast VHF/UHF channels or 12 VHF and 98 midband/superband/hyperband/ultraband cable channels.

Phase-Locked Loop Tuning. This tuning "locks-in" the signal for normal TV operation when switched to the NORM position. In the SPCL (special) position, the tuner pulls in off-frequency signals that can be received from cable antenna systems, master antenna (MATV) systems, video games, or other sources.

Automatic Degaussing Circuit. This circuit helps maintain color purity by demagnetizing the picture tube every time you turn the television on.

Electronic Power Sentry Voltage Regulator. This helps conserve energy and maintain stable voltage to the television circuits.

Special Video Filter. When switched on, this filter removes excess picture static (sometimes called "picture noise" or "snow") to give you a clearer picture, even under weak signal conditions.

Stereo Power Amplifier. This amplifier has an output of 5 watts (W) rms power per channel from 100 Hz to 10,000 Hz into 8 ohms with 2 percent or less total harmonic distortion.

Stereo Control Center. This feature includes separate bass and treble controls, a balance control to adjust the left and right levels, a hi-filter switch circuit to help eliminate high frequency noise, and a mode switch for AUX (auxiliary), STEREO, or TV AUDIO.

Four-Speaker Sound System. This system includes a 3-inch by 5-inch oval speaker and a 2½-inch round speaker in each channel for clean mid and bass frequencies.

Audio, Input/Output Jacks. Extra jacks allow you to connect optional equipment to the TV. This multifunction TV set can be used as a color TV or as a source of audio signals for hi-fi equipment. You can also use video cameras, video disc players, home video games, home computers, and videocassette recorders with your TV set. The built-in stereo power amplifier allows you to listen to stereo sound from your console stereo, stereo videodisc player, stereo videocassette recorder, and other devices equipped with stereo sound through the four TV set speakers. You can also use the audio output jacks to output your TV's sound through your console stereo.

Parental Control Pushbutton. This feature allows you to "lock-out" any channel(s) you do not want others to view.

Crosshatch Generator. This is for performing convergence and color purity adjustments.

Comb filter. Provides greater picture resolution.

Redi-Plug. This access port allows easy connection of future accessories, such as Teletext decoder and selected cable TV decoders.

VHF and UHF Antennas. These antennas are included with your set, along with a handy mounting and storage socket.

The GR-1903, 19-inch TV with remote control and all its advanced features is designed for many years of convenient entertainment and trouble-free performance, complete with broadcast TV Stereo sound.

Note: the following information provides a very detailed description of the functions of the various circuits contained in this color television receiver. By following along on the appropriate schematic diagram, you should be able to see how each component functions. Do not worry if you are not able to fully comprehend everything at one time. If you are new to the study of electronics, it will be sufficient to simply locate the components and to try to identify the sections of the main modules. With more practice, you will be amazed at how quickly you begin to see and understand much more each time you study a new electronic schematic diagram.

Tuner Module

Refer to Figs. 7-7 and 7-8 as you read the following section.

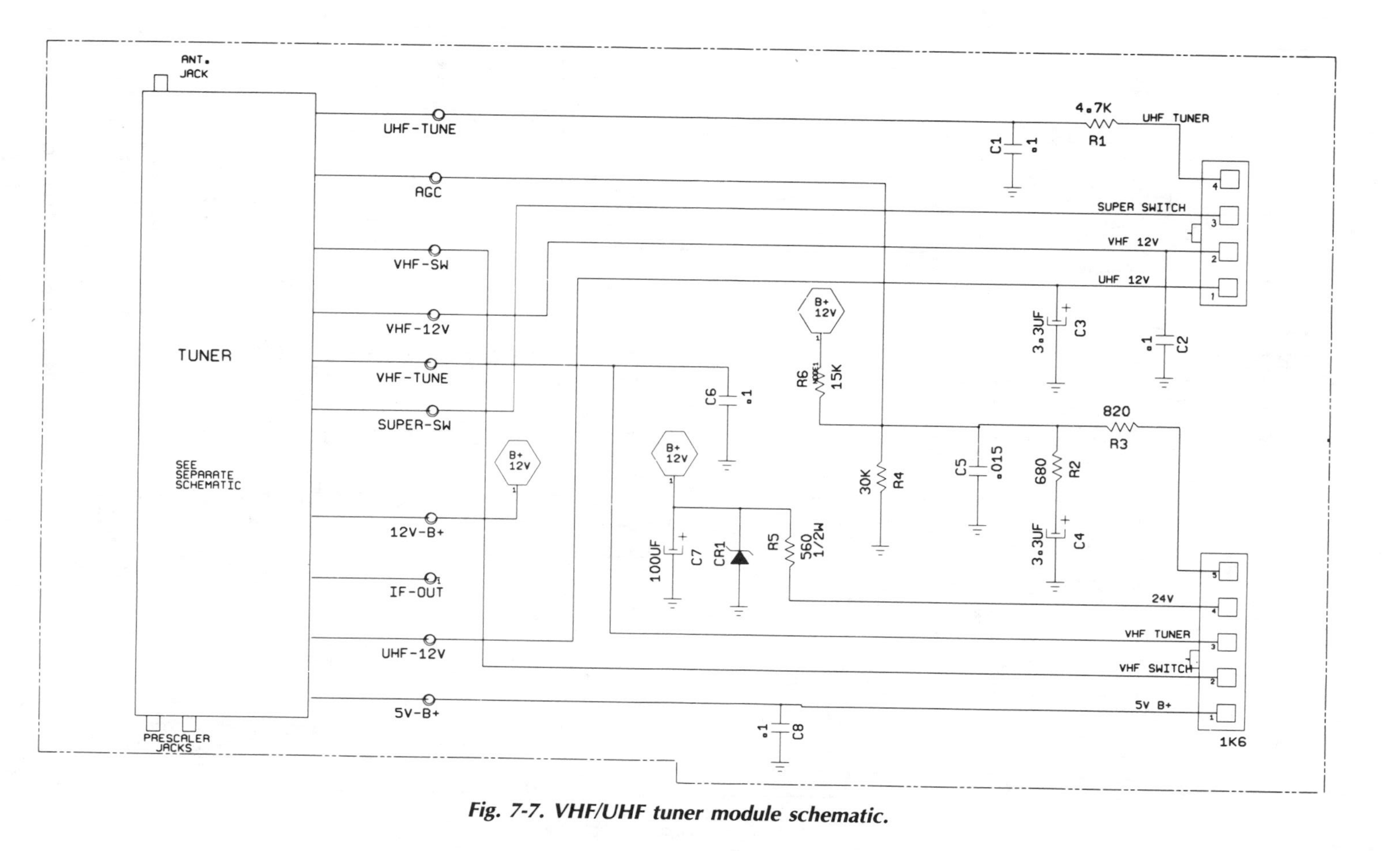

Fig. 7-7. VHF/UHF tuner module schematic.

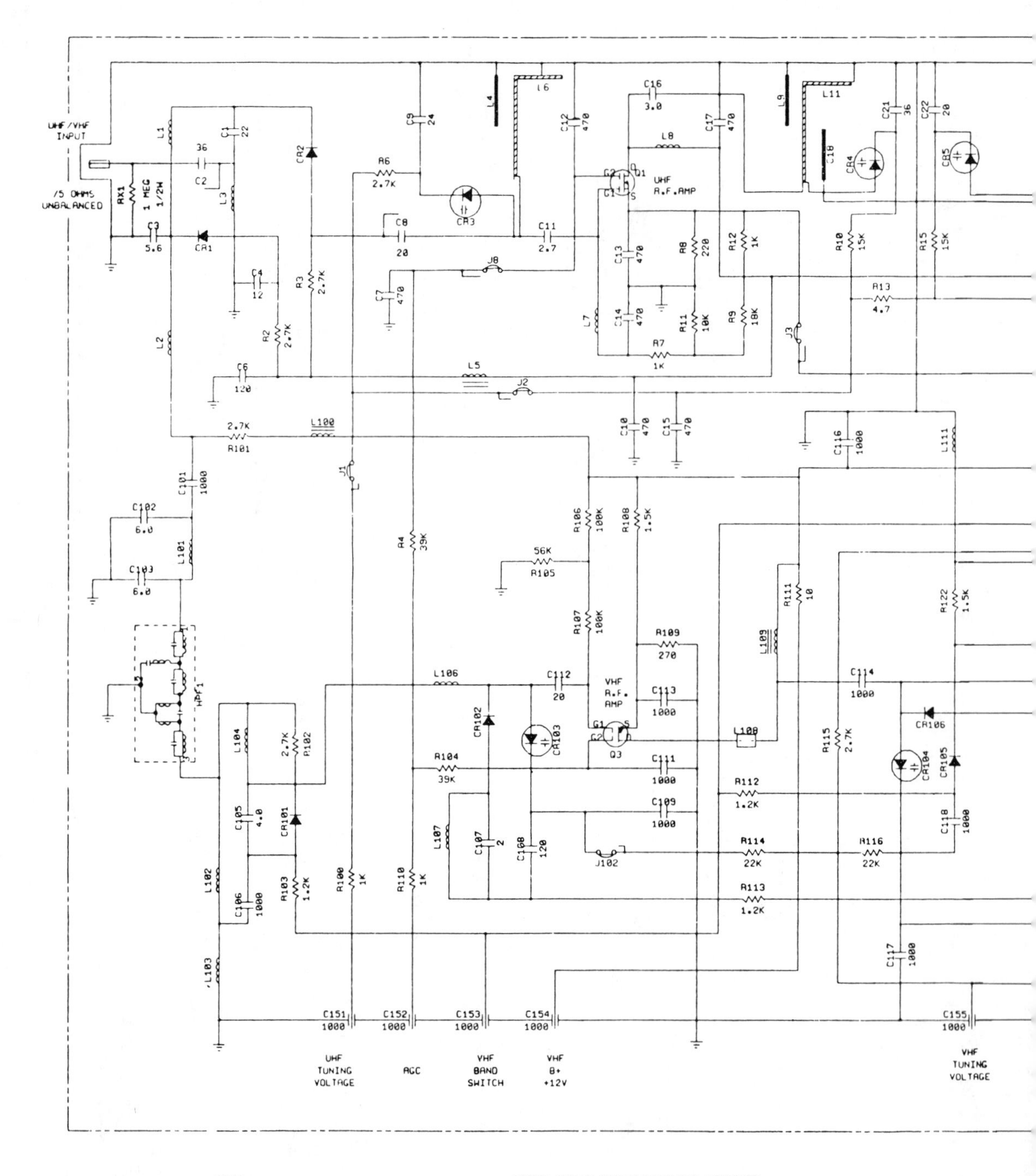

NOTES:
1. RESISTORS ARE 1/8 WATT, 5 PERCENT TOLERANCE UNLESS OTHERWISE SPECIFIED.
2. CAPACITORS ARE IN PICOFARADS UNLESS OTHERWISE SPECIFIED.

NOTES: (VALUE TAGS WITH SPECIAL MEANINGS)
A. P/L = SEE PARTS LIST FOR APPLICABLE USAGE.
B. JUMPER = JUMPER WIRE USED INSTEAD OF NORMAL PART
C. PROV = PROVISION FOR A PHYSICAL PART IN THE LAYOUT ONLY.

Fig. 7-8. VHF/UHF tuner schematic.

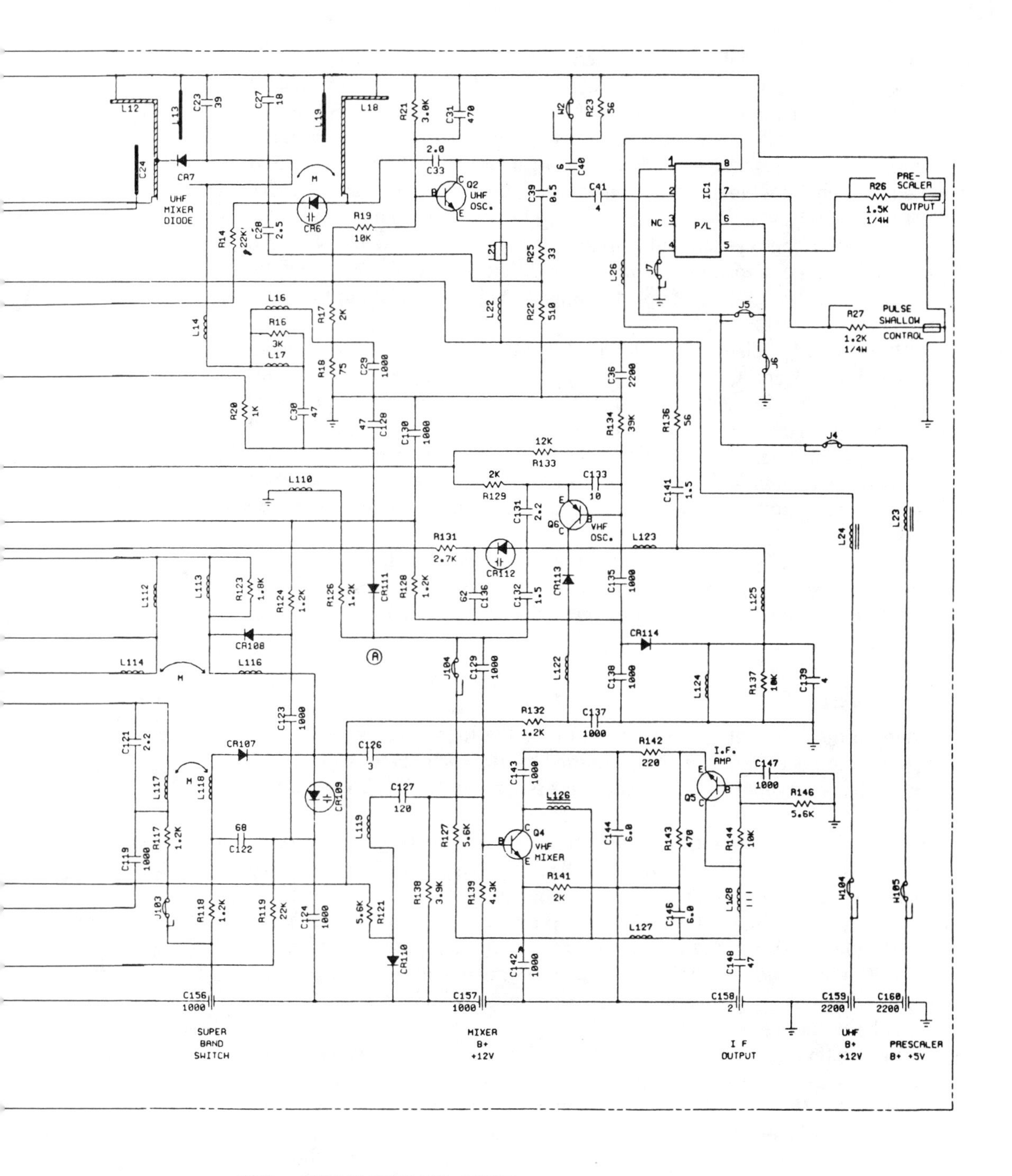

NOTES : *INDICATES CARBON COMP. RESISTOR
** INDICATES 500V COMPONENT

NOTE: C40 AND C41 ECR NO.HC-7181
C127 ECR NO.GC-7125
C136 ECR NO.FC-7105

VHF Section. The VHF signal is routed to the VHF section of the tuner by reverse-biased diodes CR1 and CR2, as seen in Fig. 7-8. After the FM band and i-f passband are filtered out, the signal is tuned by varactor diode CR103 and inductors L102, L104, L107, and L103 for low, high, super, and hyperband operation. The signal is then amplified by dual-gate MOSFET transistor Q3 and coupled through a double-tuned matching circuit consisting of varactor diodes CR104 and CR109 and inductors L112, L113, L114, and L116 to mixer transistor Q4. At this point, the selected and amplified signals are heterodyned with the signal from oscillator transistor Q6, where the difference signal is amplified by i-f amplifier transistor Q5 and coupled through tuned circuit L128 to the tuner output terminal.

UHF Section. The UHF signal is routed to the UHF tuner section by *forward*-biased diodes CR1 and CR2. It is tuned by varactor diode CR3 and inductor L6 and amplified by rf amplifier transistor Q1. From the output of Q1, the signal is coupled through double-tuned varactor/inductor circuits CR4/L11 and CR5/L12 to mixer diode CR7. The signal is mixed with the signal from oscillator transistor Q2 to produce an intermediate frequency in the 45 MHz range. This signal is amplified by transistors Q4 and Q5 and cascaded whenever the tuner is on the UHF band.

Tuning voltages, low VHF, high VHF, superband, hyperband, and UHF band switching voltages are applied to the tuner from the tuner control module. IC1, a divide-by-64 integrated circuit, divides the oscillator signal for use by the phase-locked-loop (PLL).

Tuning System

Refer to the Figs. 7-9 and 7-10 as you read the following section.

The tuning system in this TV set is formed by four main circuits: keyboard, memory, character generator, and microprocessor and interface. These circuits are described subsequently.

Keyboard. The microprocessor constantly scans the keyboard by applying negative pulses to eight output lines (KO P10 through KO P13 and KO P20 through KO P23). When you press a key on the keyboard or close one of the slide switches, the pulse returns to the microprocessor on one of the five input lines (KIN P0 through KIN P3 or KIN P30). Each of these input lines has a pull-down resistor that holds the input at a logic low when no key is being pressed. When a key is pressed, the microprocessor determines which key is closed and initiates the proper response.

Time Display. Input line 50/60 Hz is the 60 Hz input for the clock, which is displayed when the channel is changed or the ENTER button is pushed. Pull-up resistors R6774 and R6791 hold this input line at a logic high during the absence of a positive signal on the base of Q6720.

During the positive half of the 60 Hz cycle, the signal is applied to the base of transistor Q6720. This causes Q6720 to conduct and essentially grounds the junction of R6774 and R6791. Line 50/60 Hz becomes a logic low for this half cycle. During the other half cycle, Q6720 is off and line 50/60 Hz becomes logic high. The circuits inside the microprocessor use the logic-low-to-logic-high transition for the time clock.

Memory. The memory stores your favorite channels (which you program through the keyboard) so you can scan the channels without reprogramming each channel after a power interruption. Memory integrated circuit IC6800 is a special memory IC. It operates similarly to a random access memory (RAM) in that data can be stored in or retrieved from memory. However, unlike a RAM, which loses memory when power is removed, the data stored in this memory is retained for up to a year after power is removed.

Operation mode, addresses, and data are transmitted from microprocessor IC 8700 (lines C04 and C05) to memory (lines C1 and C3). Memory line C2 sets the timing of the data.

Microprocessor IC8700 supplies a two-bit address code to write data into memory. This code selects one of four 256-bit memory locations. The data is then transmitted from the microprocessor to memory, where it is stored. A START WRITE and END WRITE code is also transmitted by the microprocessor.

Microprocessor IC8700 also supplies a 20-bit address code to select the desired data to be read from memory. The memory then outputs the requested data.

To ensure against accidental writing into memory on power-up, the PCLR line (which must be logic 0 to write into memory) is pulled to +5 volts (a logic high) through resistor R6853, along with $\overline{C2}$ that is pulled up through diode CR6827. As the 5-volt (V) line rises, capacitor C6851 begins to charge through resistor R6852. Shortly after the 5-V line reaches the normal level, C6851 becomes charged enough to turn on zener diode CR6851 and transistor Q6860. When Q6860 conducts, the PCLR line becomes logic 0 and diode CR6827 is reverse-biased to isolate it from the C2 line.

Character Generator. The on-screen display is formed by rows and columns of characters. The channel/time display is two rows high and five columns wide.

The microprocessor supplies information on output lines KO P10 through KO P13 and KO P20 through KO P23 to character generator input lines DA0 through DA6 and ADM. When an LD1 input is logic low, the microprocessor outputs a seven-bit word. This word is either an address that corresponds to a row and column (for a given character) or an instruction. (The instructions would be DISPLAY FORMAT, ENABLE DISPLAY, or DISABLE DISPLAY). As LD1 changes to logic high, the seven-bit word latches into the character generator control circuits.

A second word is placed on the output lines that specifies a character on the data line required to carry out an instruction. At this time, LD1 goes back to logic low, which causes the instruction that is latched in the character generator control circuits to either be stored at the indicated memory location or be executed.

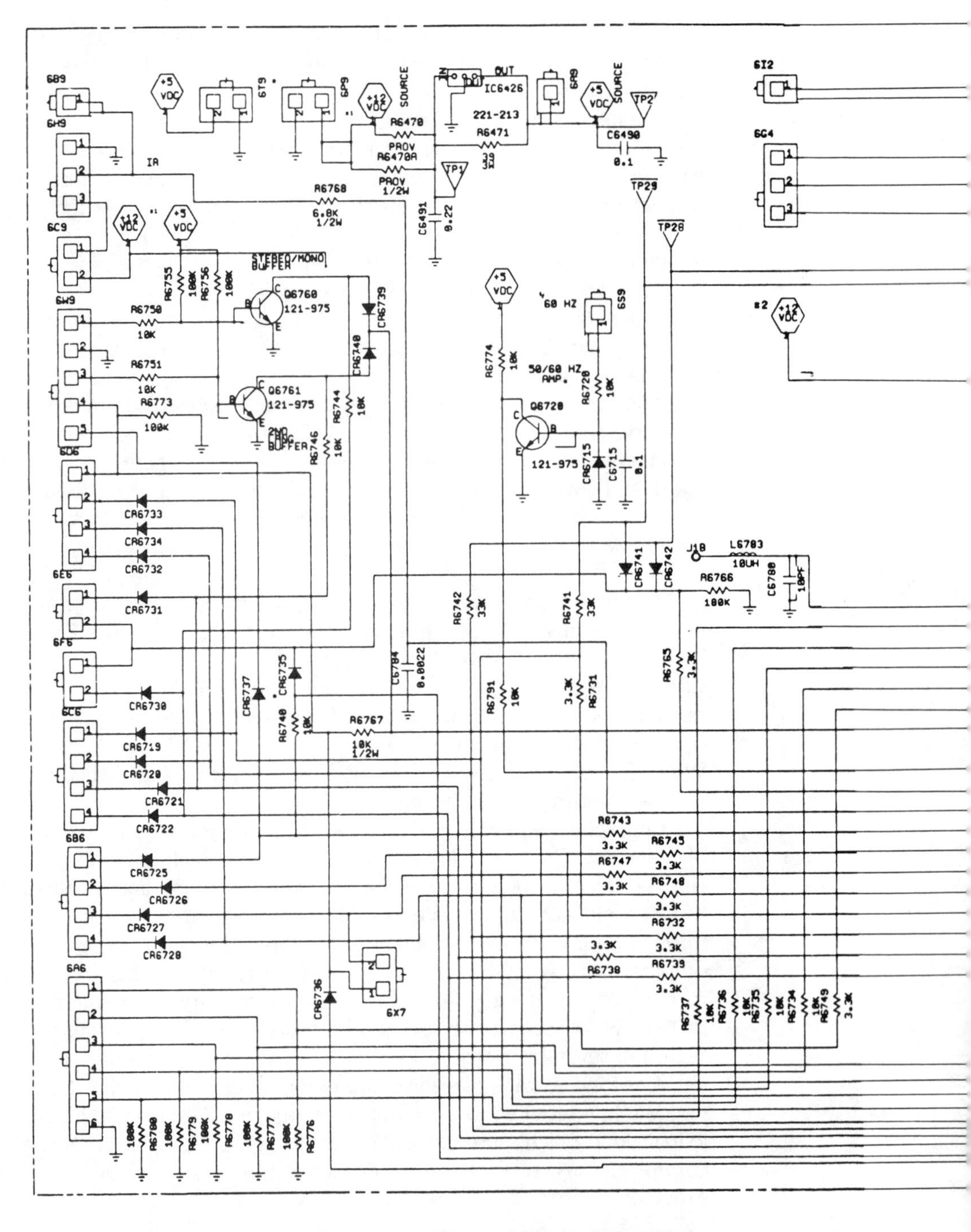

NOTES:
1. RESISTORS ARE 1/4 WATT, 5 PERCENT TOLERANCE UNLESS OTHERWISE SPECIFIED.
2. CAPACITORS ARE IN MICROFARADS UNLESS OTHERWISE SPECIFIED.

NOTES: (VALUE TAGS WITH SPECIAL MEANINGS)
A. P/L = SEE PARTS LIST FOR APPLICABLE USAGE.
B. JUMPER = JUMPER WIRE USED INSTEAD OF NORMAL PART
C. PROV = PROVISION FOR A PHYSICAL PART IN THE LAYOUT ONLY.

Fig. 7-9. Tuner control schematic (part 1).

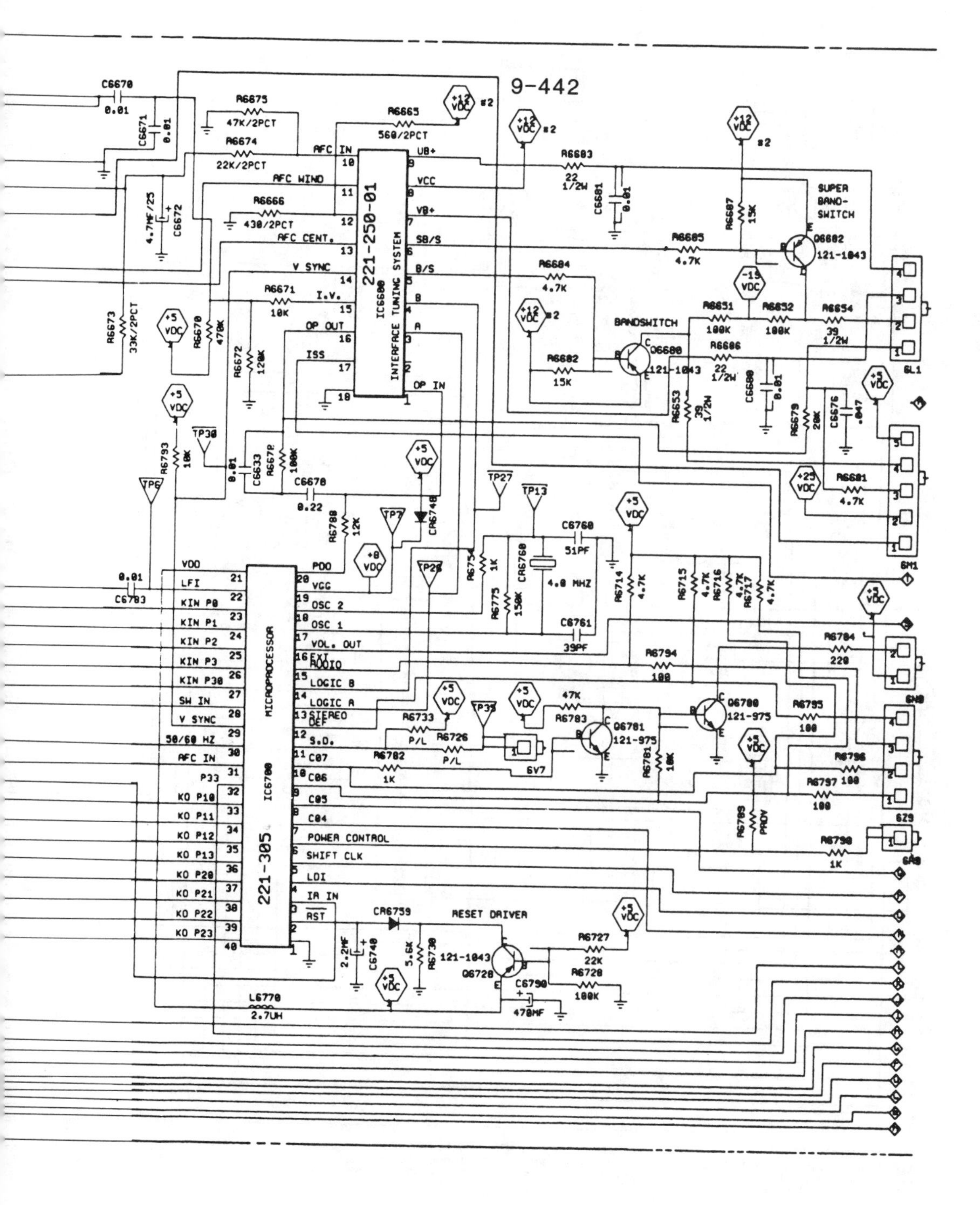
9-442
221-250-01
IC6600
INTERFACE TUNING SYSTEM
221-305
IC6700
MICROPROCESSOR
SUPER BAND-SWITCH
BANDSWITCH
RESET DRIVER
4.0 MHZ

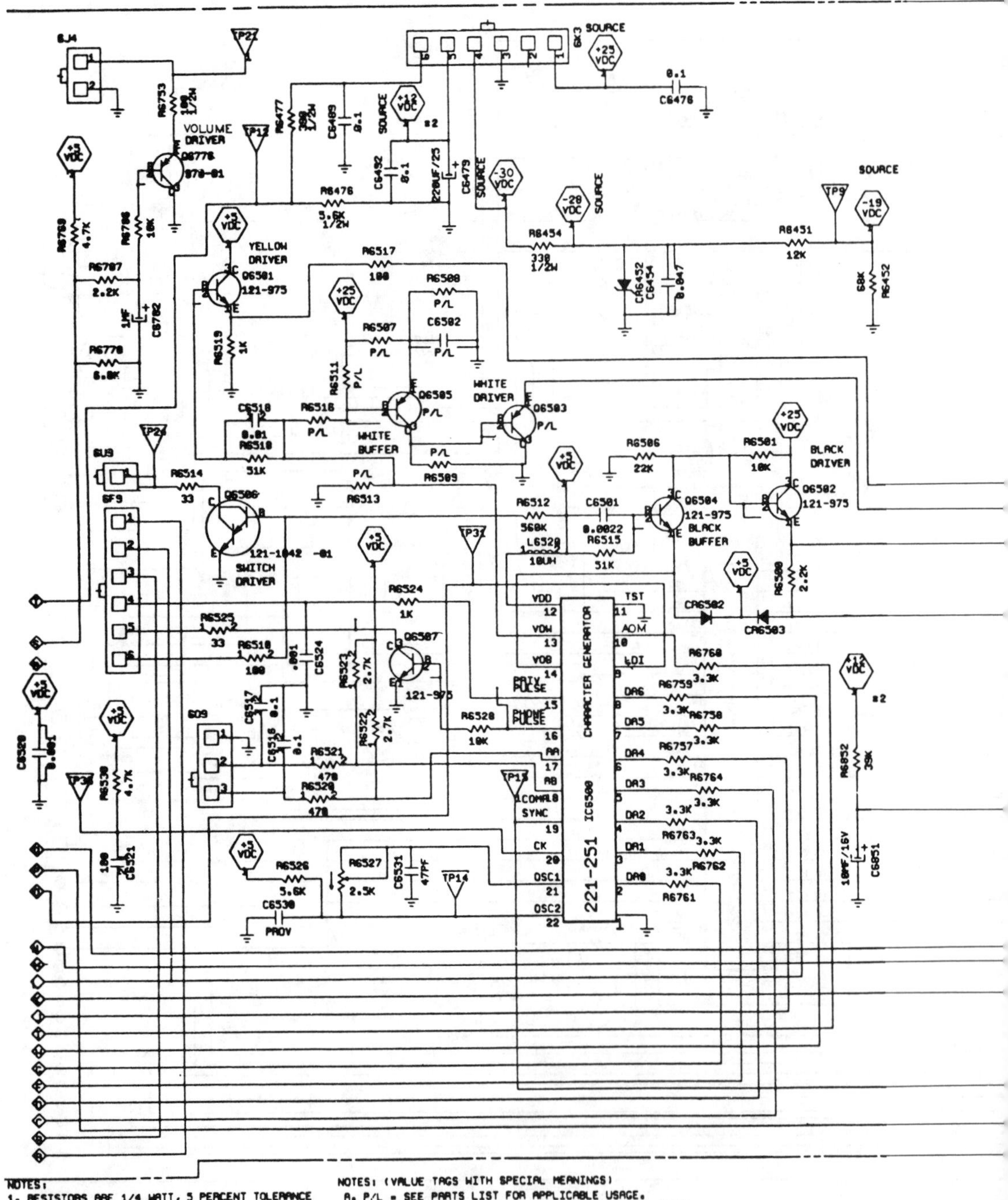

NOTES:
1. RESISTORS ARE 1/4 WATT, 5 PERCENT TOLERANCE UNLESS OTHERWISE SPECIFIED.
2. CAPACITORS ARE IN MICROFARADS UNLESS OTHERWISE SPECIFIED.

NOTES: (VALUE TAGS WITH SPECIAL MEANINGS)
A. P/L = SEE PARTS LIST FOR APPLICABLE USAGE.
B. JUMPER = JUMPER WIRE USED INSTEAD OF NORMAL PART
C. PROV = PROVISION FOR A PHYSICAL PART IN THE LAYOUT ONLY.

Fig. 7-10. Tuner control schematic (part 2).

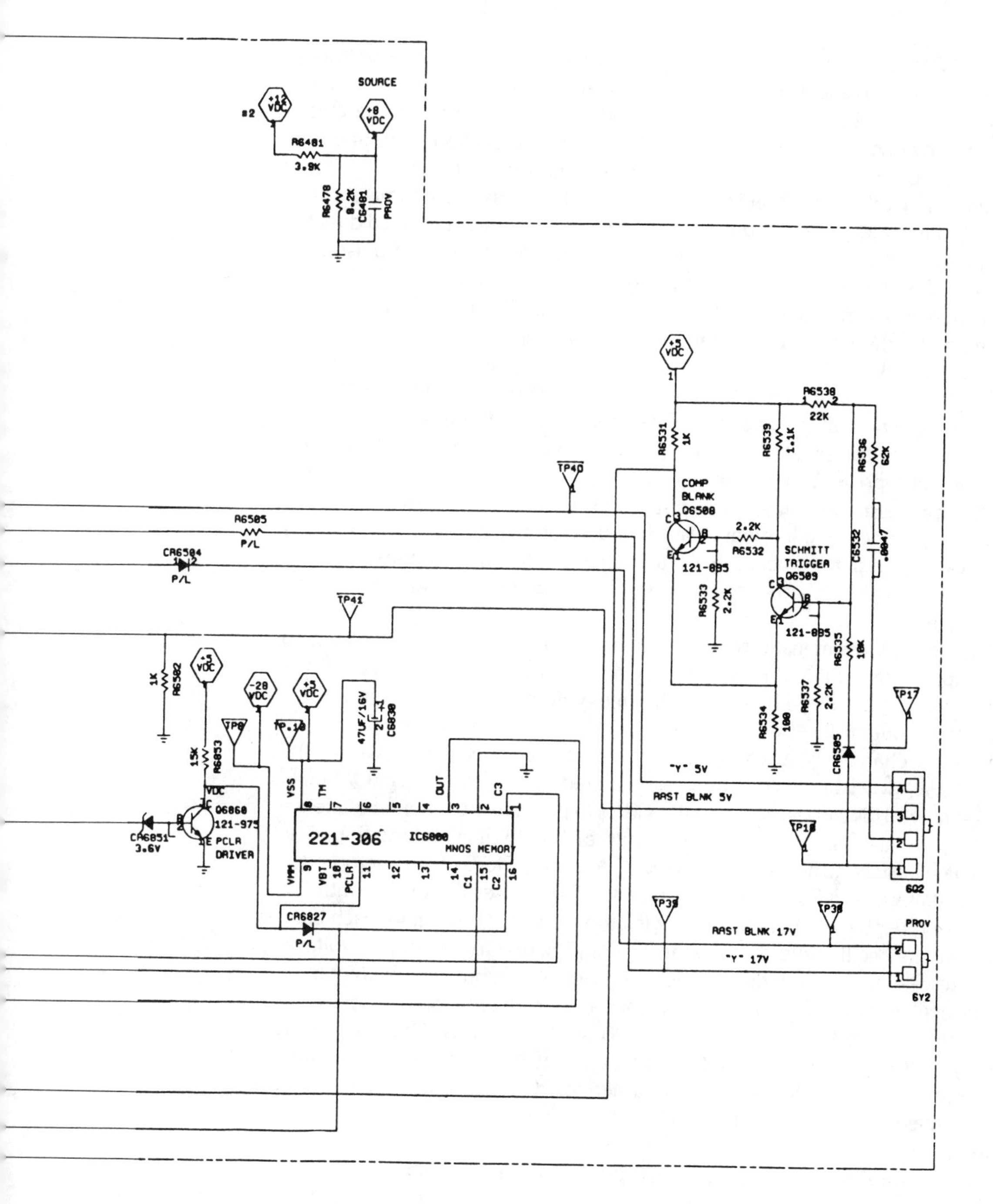
SOURCE
+12 VDC
+8 VDC
R6481
3.9K
R6478
8.2K
C6481
PROV
+5 VDC
R6538
22K
R6531
1K
R6539
1.1K
R6536
62K
TP40
COMP
BLANK
Q6508
R6505
P/L
2.2K
R6532
SCHMITT
TRIGGER
Q6509
C6532
.0047
CR6504
P/L
121-885
TP41
R6533
2.2K
R6535
10K
1K
R6502
+5 VDC
-28 VDC
+5 VDC
47UF/16V
C6830
TP8
TP.10
R6534
100
R6537
2.2K
TP17
CR6505
15K
R6853
"Y" 5V
RAST BLNK 5V
VSS
TM
OUT
C3
Q6860
121-975
PCLR
DRIVER
CR6851
3.6V
221-306
IC6800
MNOS MEMORY
TP18
VMM
VBT
PCLR
C1
C2
6Q2
TP39
TP38
CR6827
P/L
RAST BLNK 17V
PROV
"Y" 17V
6Y2

The ADM input on the character generator IC is used for address mode control. This input is a logic low during the time the preceding operation occurs. If ADM is a logic high, the previously latched address in the control circuitry increments by one. The only data for this new memory location is supplied by the microprocessor and is latched by a positive transition of LDI.

An RC network at OSC1 and OSC2 allows you to adjust the internal 4 MHz clock. The character generator utilizes a composite sync-pulse input fed directly to an input on the microprocessor. The microprocessor breaks the pulse into the required horizontal and vertical outputs. Horizontal information is derived from a counter, which places the display box on the CRT screen in the location determined by the setting of the set-up potentiometer.

Outputs VOW and VOB provide the on-screen display. A logic low output at VOB provides a black background, and a logic high output at VOW provides yellow characters. Buffer and driver transistors produce the proper amplitude and polarity. The drivers also interface to the main module to override the video so the appropriate information is displayed on the screen.

Microprocessor and Interface. Crystal CR6760, between the two OSC pins, sets the frequency of the internal 7.16 MHz oscillator. This oscillator provides the clock required for operation of the microprocessor and the reference frequency for the phase-locked-loop (PLL) circuits contained in the microprocessor.

The serial bit stream that is generated in the infrared preamp circuits is applied to the input IR IN, where it is decoded by the microprocessor. The VOL OUT output provides dc voltage in 64 steps, through volume driver transistor Q6776, to adjust the output level of the audio IC.

When a channel is selected, the microprocessor checks the state of the CATV/NORMAL switch and then obtains the required divide ratio for its ROM (read only memory). This causes the internal dividers to produce the required reference frequency for the PLL along with two bits of bandswitch information. The bandswitch logic is latched at outputs LOGIC A and LOGIC B of the microprocessor IC and at B/S and SB/S of the interface IC. This causes the tuner control to select the proper band. Minimum tuning (clamp) voltage is also set for each band to prevent the oscillator in the tuner from stopping due to low tuning voltage.

A prescaler in the tuner divides the tuner oscillator frequencies by 65 before they are applied to the LFI input of the microprocessor. The microprocessor further divides this frequency and then applies it to the PLL circuits. The PLL adjusts the duty cycle of a 2 kHz, 5-V peak-to-peak output at PDO. This signal is applied through resistor R6788 to OP IN on the interface IC. An operational amplifier that is connected between OP IN and OP OUT is supplied by a 32-V, 13.6-milliampere (mA) constant current supply at ISS.

Resistor R6678 and capacitor C6678 provide feedback from the OP OUT to the OP IN for filtering. The filtered dc tuning voltage at OP OUT is applied to the tuner and is adjusted to obtain and maintain phase lock between the oscillator and microprocessor-supplied reference frequency.

In the normal mode, the tuning system always places the tuner oscillator at the correct frequency to receive an FCC-specified channel frequency. At some CATV or MATV installations, the channel oscillator frequencies are offset from the FCC-designated standard. This is done purposely to lessen interference. Some inexpensive TV games also operate on offset frequencies. Provision is made to capture channels that are offset by as much as 3.25 MHz. This feature is activated whenever the AFC switch is in the SPECIAL position.

Pull-down resistor R6766 holds the AFC IN input to the microprocessor at logic low when a logic high is not present at the anode of diodes CR6741 or CR6742. AFC voltage from the i-f module is applied to the interface IC at AFC IN. Internally, this voltage is applied to three comparators. The outputs from two of these comparators are connected to the AFC WIND output and produce a logic high if the voltage is in the window (+1 V to +5 V). The output of the third comparator is applied to the AFC CENT and produces a logic high or logic low, depending on which side of the center voltage (+3 V) the AFC happens to be. When a channel is selected in the special AFC mode, the system first synthesizes the correct frequency for that channel.

The microprocessor places a logic low on the KO P21 output, causing the anode of diode CR6742 to become clamped to logic low through resistor R6742. This happens regardless of the AFC window level. The microprocessor now checks input AFC IN. A logic high from the AFC CENT output of the interface IC, applied through diode CR6741, indicates the need for tuning down. A logic low indicates the need for tuning up. The microprocessor steps the tuning voltage in small increments which, in turn, change the tuner oscillator frequency. If the tuning voltage is being stepped down, the microprocessor looks for either a high-to-low transition of the AFC CENT output at AFC IN or looks for a low-to-high transition if it is stepping up. This transition indicates that the center frequency has been found and the stepping ceases.

At this time, the processor checks for the validity of the carrier. For a carrier to be valid, the output of the AFC window must be a logic high, and a vertical pulse train must be present at microprocessor input V SYNC. Vertical sync from the sync integrator on the main module is applied to input I.V. on the interface IC. Here, it passes through a comparator to the V SYNC output. If the carrier is invalid, stepping resumes.

After each tuning voltage step, if the center frequency is not found, output KO P21 goes to a logic high and KO P20 goes to a logic low. This applies the AFC window signal input AFC IN, which disables AFC CENT. If AFC WIND goes to a logic low, the microprocessor determines that the edge of the AFC window is reached and begins stepping in the opposite direction. If the microprocessor does not find a valid carrier, stepping continues to the other end of the window, where the AFC WIND output again becomes a logic low and stops the stepping action. During the time the system is looking for a valid carrier, the raster (rectangle of light projected when no signal is present) is blanked and the sound is muted. When a valid carrier is located, or if the system searches the entire window without finding one, blanking and mute are disabled.

When the STEREO,2ND AUDIO, or MONO key is pressed, the negative pulse from microprocessor KO P23 is returned to KIN P1. The microprocessor then tells the stereo decoder that stereo has been selected and turns it on. If the stereo decoder detects the presence of the pilot signal, a logic 0 is entered on the base of Q6760 to forward-bias CR6739 and activate microprocessor SW IN. The microprocessor places a logic 0 at Q6780 and turns on the stereo LED indicator. If no stereo pilot signal is detected, the microprocessor turns the stereo LED off via Q6780 and Q6781.

If the second audio, or language, is detected, the microprocessor is informed in the same manner as the stereo pilot to activate Q6761 and forward-bias CR6740 and the microprocessor. The stereo decoder is thus informed and the second language function is activated.

I-f Amplifier

Refer to Figs. 7-11 and 7-12 as you read the following sections.

The i-f amplifier, detector, and AGC circuitry stages shown in Fig. 7-11 are connected internally in the integrated circuit IC1201.

The signal from the tuner passes through a matching pad, C1226 and L1226, and then to surface wave filter SWIF. Because the input impedance of the first i-f stage is almost purely resistive, it is similar to having the output of the SWIF tuned. The result is that less power is lost in the interface between the SWIF and the first i-f stage.

The i-f amplifier consists of five stages of amplification (all of the differential type). The first stage is a grounded base stage to produce the low impedance input. This stage has no AGC applied to it.

The second and third stages are conventional differential amplifiers. They contain variable emitter degeneration to produce about 23 dB of gain reduction per stage. This variable emitter degeneration is brought about by controlling the forward bias on a diode which parallels the emitter resistor.

The fourth stage is the peaker stage. This stage contains a high-Q, series-tuned circuit in the form of a ceramic filter U1203 (called a *piezonator*) selectively placed across the emitter circuit. This selection is controlled by the AGC voltage acting on a forward-biased diode, similar to the previous AGC stages. When this high-Q circuit (which is tuned to the picture carrier) is placed across the emitter, it raises the gain of the stage to the picture carrier, but leaves the gain of the rest of the i-f band a constant. The gain to the picture carrier is designed to increase 6 dB at maximum gain, placing the picture carrier at the top of the i-f response curve. The rest of the response will remain the same as the strong signal condition.

When the picture carrier changes 6 dB as a function of the AGC voltage, it amounts to 6 dB more gain reduction in the amplifier. This gain reduction, along with the 46 dB gain reduction of the two previous stages, adds to a total gain reduction of 52 dB.

The fifth stage is the output stage. Although the five stages are operated differently, only one output is used in this stage.

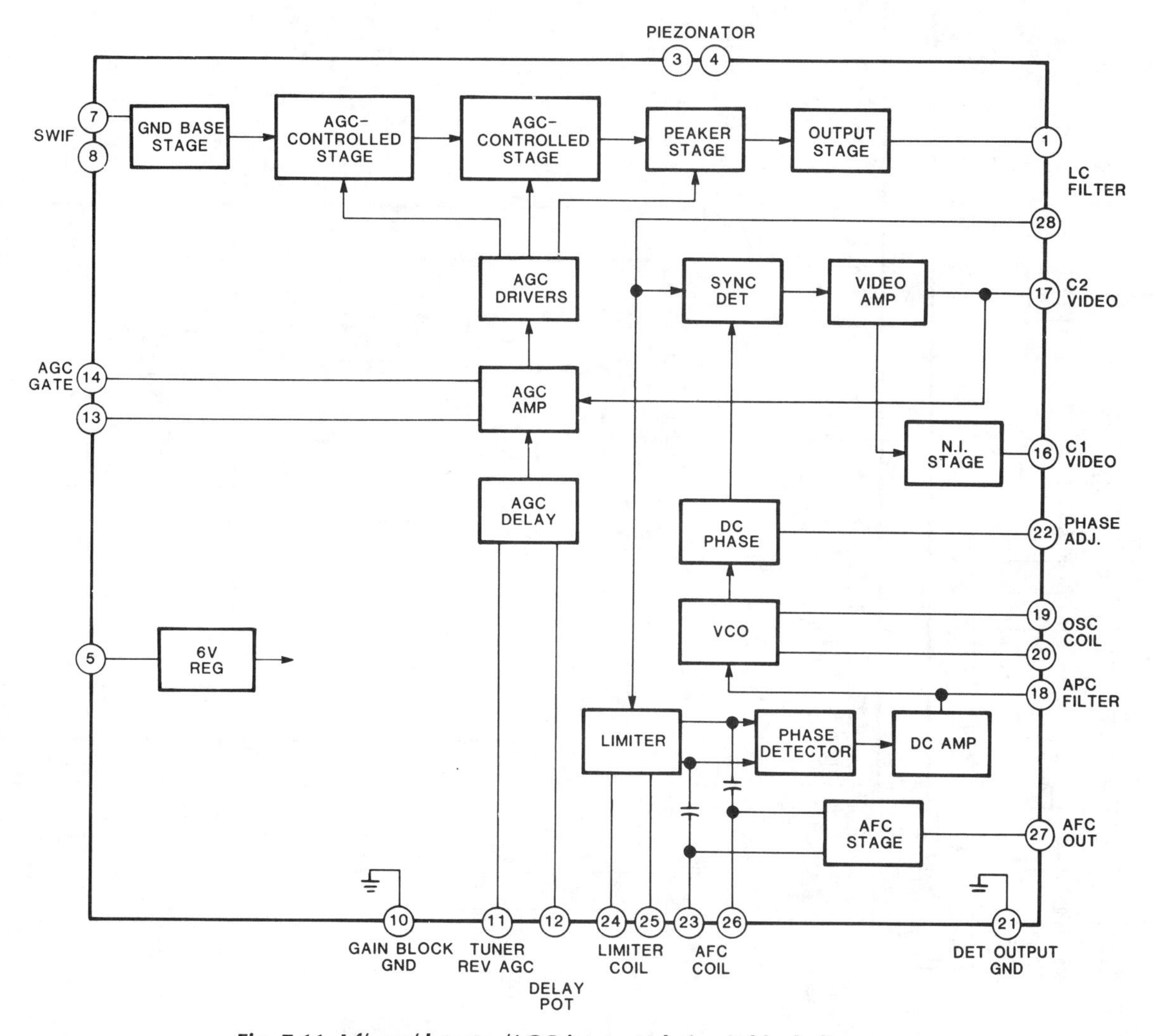

Fig. 7-11. I-f/amp/detector/AGC integrated circuit block diagram.

The signal from the output amplifier stage, at pin 1 of IC1201, is coupled through capacitors C1210 and C1212 to pin 28 where it is split two ways; one output goes to the synchronous detector stage; the other goes to the limiter stage.

The limiter stage collector circuit contains back-to-back diodes as the main limiting element. The limiter signal is sent to the phase detector. In addition, a part of the limiter output is used to drive the AFC circuitry. The signal that is applied to the phase detector is multiplied by its other input, which becomes the reference oscillator signal. When these two signals are 90 degrees out of phase, the output of the multiplier is zero. At any other phase condition, the output will produce a beat note accompanied by a dc component. This signal is filtered, amplified, and then used to control the VCO (voltage controlled oscillator) reference oscillator.

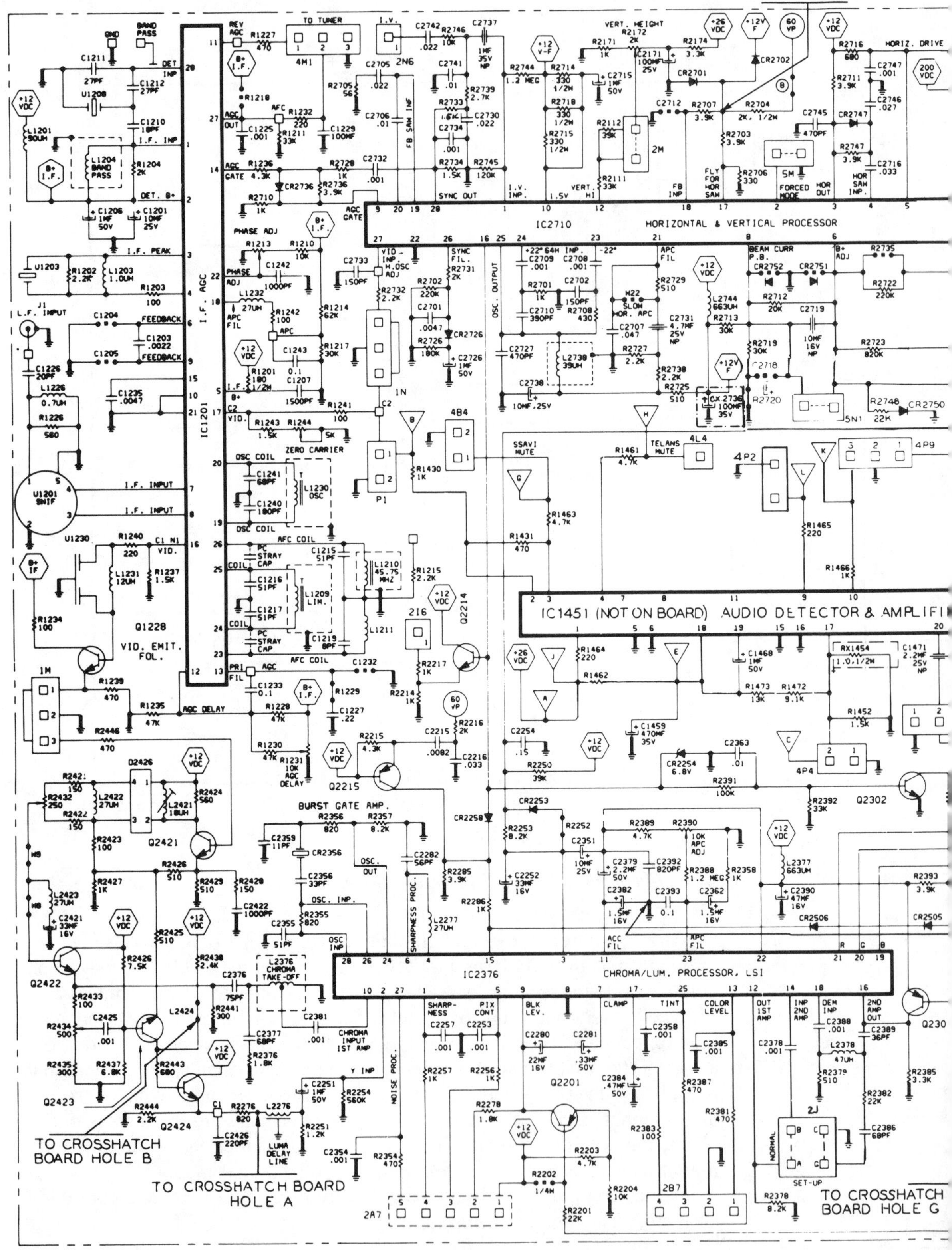

Fig. 7-12. Main module schematic.

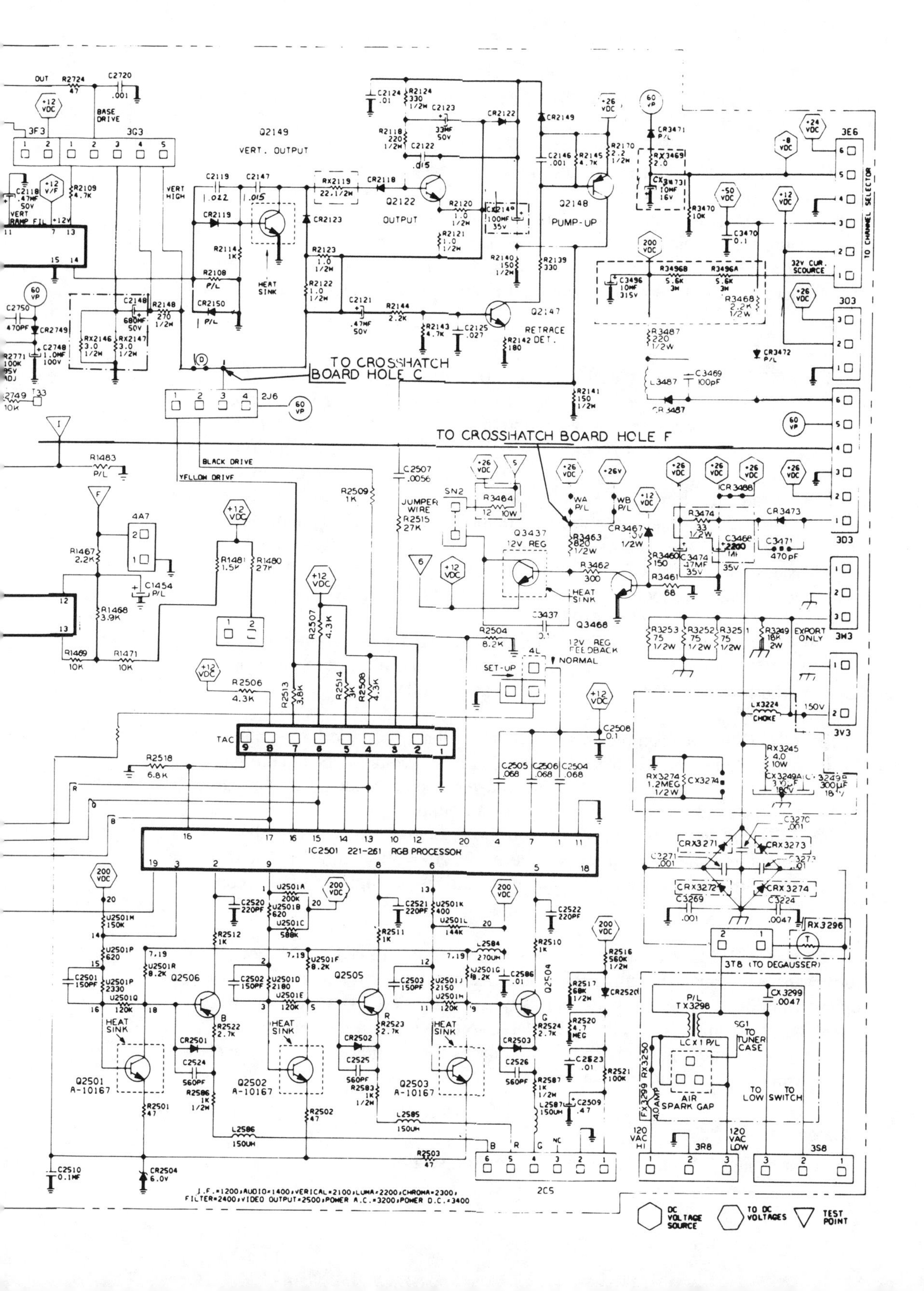

OUT
R2724
47
C2720
.001
BASE DRIVE
3F3
3G3
Q2149
VERT. OUTPUT
VERT HIGH
C2119
.022
C2147
.015
CR2119
R2114
1K
HEAT SINK
R2108
P/L
CR2150
P/L
C2148
680MF
50V
R2148
270
1/2W
RX2146
3.0
1/2W
RX2147
3.0
1/2W
C2750
470PF
CR2749
C2748
1.0MF
100V
R2771
100K
T33
C2124
.01
R2124
330
1/2W
C2123
33MF
50V
R2118
220
1/2W
C2122
.015
RX2119
22.1/2W
CR2118
Q2122
OUTPUT
CR2123
R2120
1.0
1/2W
R2121
1.0
1/2W
R2123
1.0
1/2W
R2122
1.0
1/2W
C2121
.47MF
50V
R2144
2.2K
R2143
4.7K
C2125
.027
CR2122
CR2149
C2146
.001
R2145
4.7K
R2170
2.2
1/2W
Q2148
PUMP-UP
R2140
150
1/2W
R2139
330
Q2147
RETRACE DET.
R2142
180
R2141
150
1/2W
TO CROSSHATCH BOARD HOLE C
2J6
60 VP
CR3471
P/L
RX3469
2.0
10MF
16V
R3470
10K
C3470
0.1
C3496
10MF
315V
R3496B
5.6K
3W
R3496A
5.6K
3W
R3468
2.2K
1/2W
32V CUR. SCOURCE
3E6
TO CHANNEL SELECTOR
303
R3487
220
1/2W
L3487
C3469
100pF
CR3487
CR3472
P/L
TO CROSSHATCH BOARD HOLE F
R1483
P/L
BLACK DRIVE
YELLOW DRIVE
4A7
R1467
2.2K
C1454
P/L
R1468
3.9K
R1469
10K
R1471
10K
R1481
1.5K
R1480
27K
R2509
1K
C2507
.0056
JUMPER WIRE
R2515
27K
SN2
R3484
12
10W
Q3437
12V REG
WA
P/L
WB
P/L
R3463
820
1/2W
CR3467
10V
1/2W
R3462
300
HEAT SINK
Q3468
C3437
0.1
R3460
150
R3461
68
R3474
33
1/2W
C3474
47MF
35V
C3468
2200
35V
CR3473
C3471
470pF
ICR3488
R2504
8.2K
12V REG FEEDBACK
NORMAL
SET-UP
4L
3M3
R3253
75
1/2W
R3252
75
1/2W
R3251
75
1/2W
R3249
18K
2W
EXPORT ONLY
LX3224
CHOKE
150V
3V3
RX3245
4.0
10W
RX3274
1.2MEG
1/2W
CX3274
CX3249A
CX3249B
300µF
R2506
4.3K
R2513
3.8K
R2507
4.3K
R2514
3K
R2508
4.3K
TAC
R2518
6.8K
C2508
0.1
C2505
.068
C2506
.068
C2504
.068
IC2501 221-261 RGB PROCESSOR
C3270
.001
CRX3271
CRX3273
C3271
.001
C3273
.01
CRX3272
CRX3274
C3269
.001
C3224
.0047
RX3296
3T8 (TO DEGAUSSER)
U2501M
150K
U2501P
620
C2501
150PF
U2501P
2330
U2501R
8.2K
U2501Q
120K
Q2506
C2520
220PF
R2512
1K
U2501A
200K
U2501B
620
U2501C
588K
C2502
150PF
U2501D
2180
U2501E
120K
U2501F
8.2K
Q2505
R2511
1K
C2521
220PF
U2501K
400
U2501L
144K
C2503
150PF
U2501J
2150
U2501H
120K
U2501G
8.2K
L2584
270UH
C2586
.01
C2522
220PF
R2510
1K
Q2504
R2516
560K
1/2W
R2517
68K
1/2W
CR2520
R2520
4.7
MEG
C2523
.01
R2521
100K
C2509
.47
R2587
1K
1/2W
L2587
150UH
R2522
2.7K
R2523
2.7K
R2524
2.7K
CR2501
CR2502
CR2503
C2524
560PF
C2525
560PF
C2526
560PF
R2586
1K
1/2W
R2583
1K
1/2W
Q2501
A-10167
Q2502
A-10167
Q2503
A-10167
R2501
47
R2502
47
L2586
150UH
L2585
150UH
R2503
47
C2510
0.1MF
CR2504
6.0V
2C5
CX3299
.0047
P/L
TX3298
SG1
TO TUNER CASE
LCX1 P/L
AIR SPARK GAP
RX3250
FX3299
4.0AMP
TO LOW
TO SWITCH
120 VAC HI
120 VAC LOW
3R8
3S8
I.F.=1200;AUDIO=1400;VERICAL=2100;LUMA=2200;CHROMA=2300;
FILTER=2400;VIDEO OUTPUT=2500;POWER A.C.=3200;POWER D.C.=3400
DC VOLTAGE SOURCE
TO DC VOLTAGES
TEST POINT

The VCO signal is applied to the synchronous detector stage through a voltage-controlled phase shift stage. The variable phase shift control R1213, allows the sync detector to be adjusted without affecting the limiter adjustment.

The video amplifier stage provides the correct levels for the C1 and C2 video outputs at pins 16 and 17 of the IC1201. The C1 signal is inverted and fed to output pin 16. This action takes place after it overcomes an offset voltage. The C2 output is an emitter follower and is output on pin 17.

A 6-V regulator functions as a shunt regulator, using a 5.3-V zener diode as the reference unit to power the internal circuitry of IC1201.

AGC System

The AGC function is included in the IC1201 circuitry on the main module. However, the sync separator function is included with the horizontal/vertical processor IC2710. Because of the separation of the sync/AGC into separate ICs, an external circuit is used to provide the function of AGC gate generation.

Sync pulses from the sync output at pin 28 of IC2710, are ac-coupled to the flyback gate signal from pin 9. When the sync and flyback gate signals are in phase, the AGC gate signal at the anode of the diode CR2736 is mainly sync. However, when the sync pulses are absent, the AGC gate signal is a reduced flyback gate signal. Diode CR2736 attenuates the noise during the non-flyback gate period. From the anode of diode CR2736, the AGC gate signal is coupled through R1236 to pin 14 of IC1201.

The functions of the i-f amplifiers, synchronous detector, AGC detector, AGC delay, i-f AGC amplifier, and the rf AGC amplifier are all incorporated into IC1201. The functions of the AGC detector are as follows:

- AGC Detector—Samples the video during the sync period and charges or discharges the AGC filter capacitor C1227 to a voltage which represents the signal strength. The AGC filter voltage is applied to the i-f and through the AGC delay circuit to the rf AGC amplifiers.
- AGC Delay—Determines the threshold below which only the i-f amplifier's gain is adjusted by the AGC. Above this threshold, the rf amplifier is adjusted for maximum gain, and the i-f amplifier is being controlled.

Capacitor C1233 is connected between pins 12 and 13 of IC1201. For abrupt changes in signal strengths above the AGC threshold, this capacitor causes the i-f amplifier to alter its gain and stabilize the video signal. During this transition, the rf amplifier's gain is not altered. After a time, the i-f amplifier will return to its minimum gain and the rf amplifier's gain will adjust to a new level which will correspond to the new signal strength.

Stereo I-f Module

Refer to Fig. 7-13 as you read the following section.

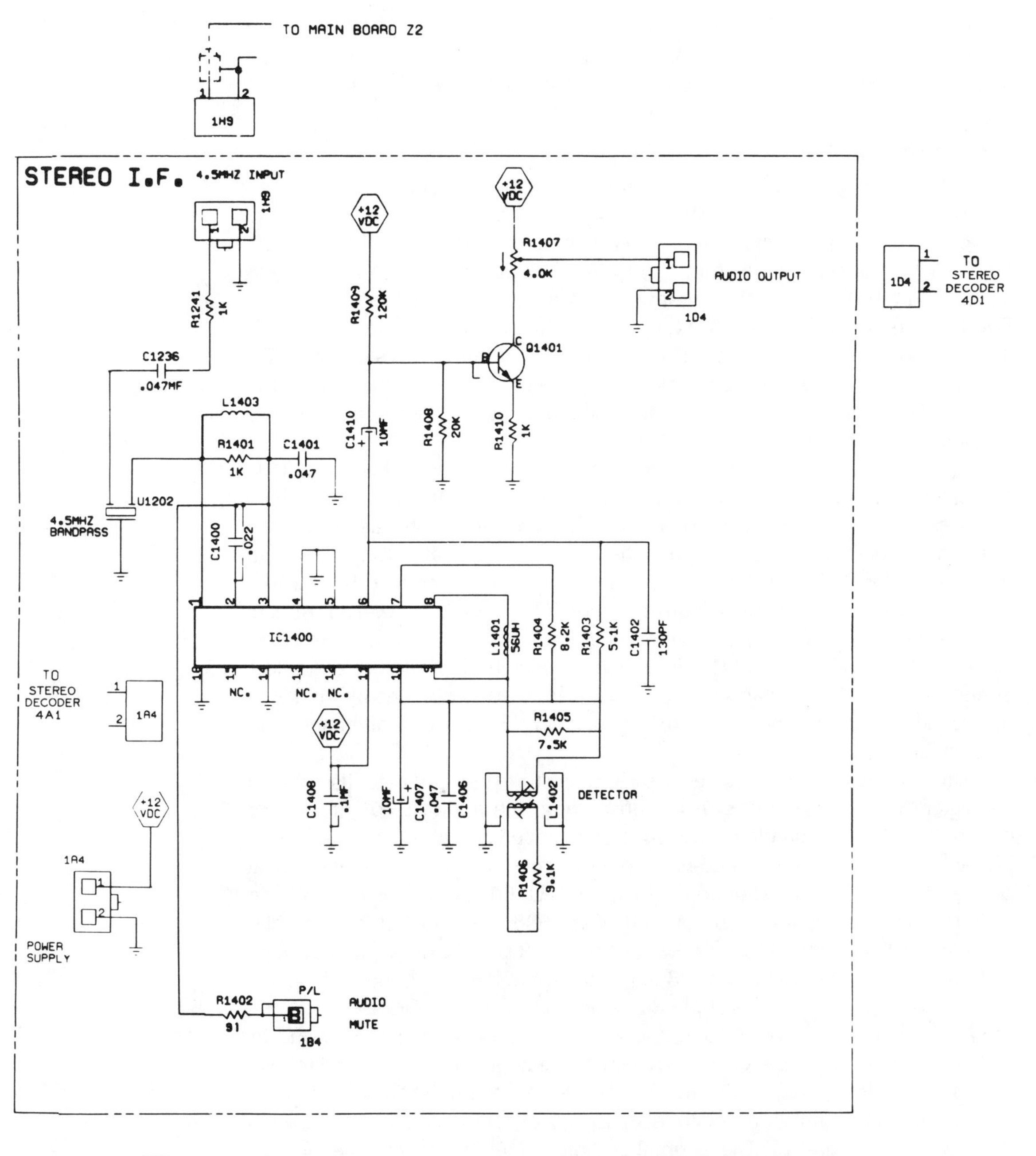

NOTES:
1. RESISTORS ARE 1/4 WATT, 5 PERCENT TOLERANCE UNLESS OTHERWISE SPECIFIED.
2. CAPACITORS ARE IN MICROFARADS UNLESS OTHERWISE SPECIFIED.

NOTES: (VALUE TAGS WITH SPECIAL MEANINGS)
A. P/L = SEE PARTS LIST FOR APPLICABLE USAGE.
B. JUMPER = JUMPER WIRE USED INSTEAD OF NORMAL PART
C. PROV = PROVISION FOR A PHYSICAL PART IN THE LAYOUT ONLY.

Fig. 7-13. Stereo i-f schematic.

Integrated circuit IC1400 employs double-tuned detector coil L1402 for better linearity in demodulating the 4.5 MHz signal. The output audio signal is applied to the base of Q1401 and exits at connector 1D4 to the stereo decoder module.

Stereo Decoder Module

Refer to the Fig. 7-14 as you read the following section.

The composite audio input signal, which consists of the stereo and second audio program (SAP), is applied to the base of emitter-follower transistor Q1625. Signals which appear at the emitter of Q1625 are coupled to IC1601 through R1601 and C1602. IC1601 is the same type of stereo decoder used in standard FM receivers with some slight modifications. The output from pins 4 and 5 of IC1601 contain L+R and L−R signals. The output at pin 7 drives a stereo LED indicator when the pilot stereo signal is detected. A1604 is a 78.67 kHz trap which attenuates the second audio and reduces crosstalk between the stereo and second audio.

R1611 adjusts the free-running frequency of the phase-locked-loop (PLL) contained in IC1601. IC1601 also contains a stereo switch and a decoder which provides L+R and L−R at its output. The L+R and L−R signals are passed through low-pass filters A1602 and A1603. The L+R signal is routed through matrix decoder R1619, R1620, R1603, and R1607 and amplified by IC1609. The L−R signal is passed onto the dbx decoder. R1618 is a L−R level adjustment to correct for gain differences in stereo decoder IC1601. The L−R signal then passes through Q1617 and Q1618, which are electronic switches, and allow either second audio programs or the L−R audio to go to dbx decoder IC1607. R1703 adjusts the timing of the dbx circuit and R1709 is a highband trim adjustment.

The purpose of the dbx decoder is to restore the audio to its original characteristics (noise reduction). In transmitting, the audio is compressed or encoded, and then expanded or decoded at the receiver. This process is accomplished to enhance the signal-to-noise performance. From the dbx decoder at pin 15, the signal is passed on to dual-amplifier U1608 at pin 2 via R1769 and C1720. The signal is amplified by half of IC1608 and then returned to dbx decoder pin 5 through C1707 and R1704. The L−R signal exits at pin 8 through C1618 to the second half of U1608 at pin 5, where it is amplified and exits at pin 7. The amplified signal is applied to matrix amplifier IC1609 together with the L+R audio signal from C1610 and R1603. In the matrix amplifier and its associated circuits, the L+R and L−R signals are added and subtracted to give the L and R audio signals. R1626 and R1627 are the left and right matrix controls.

The left and right audio channels then go to four-channel stereo switch IC1603 at pins 13 and 11. The external audio channels are also applied to pins 2 and 4 of the stereo switch. When the SECOND AUDIO PROGRAM signal is selected, it is switched by Q1618 and applied to IC1603 pins 11 and 13 instead of the left and right audio channels. Audio inputs to U1603 are controlled either by dc voltages applied to connector 4B6, or electronically via Q1616 or Q1618.

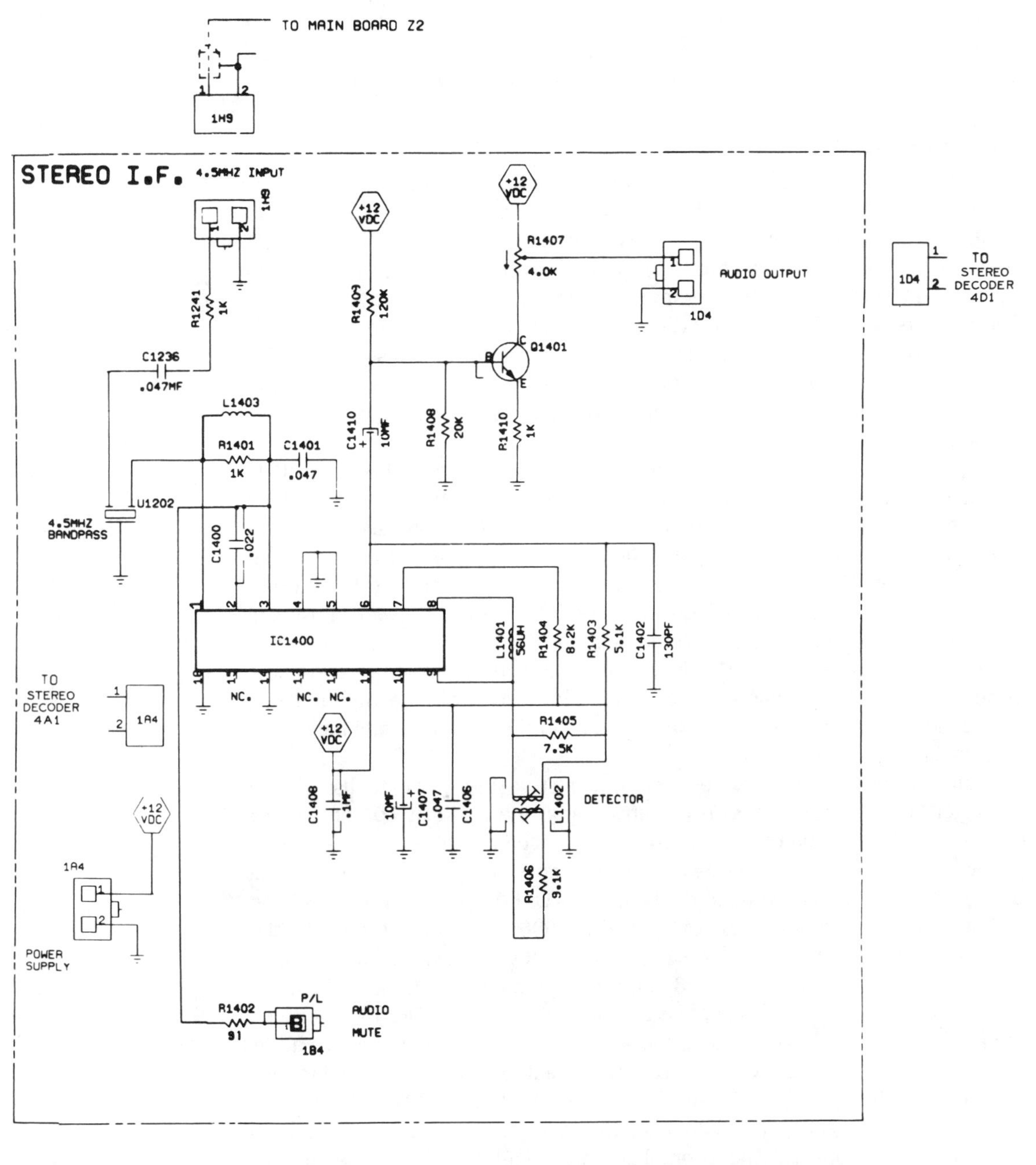

NOTES:
1. RESISTORS ARE 1/4 WATT, 5 PERCENT TOLERANCE UNLESS OTHERWISE SPECIFIED.
2. CAPACITORS ARE IN MICROFARADS UNLESS OTHERWISE SPECIFIED.

NOTES: (VALUE TAGS WITH SPECIAL MEANINGS)
A. P/L = SEE PARTS LIST FOR APPLICABLE USAGE.
B. JUMPER = JUMPER WIRE USED INSTEAD OF NORMAL PART
C. PROV = PROVISION FOR A PHYSICAL PART IN THE LAYOUT ONLY.

Fig. 7-13. Stereo i-f schematic.

Integrated circuit IC1400 employs double-tuned detector coil L1402 for better linearity in demodulating the 4.5 MHz signal. The output audio signal is applied to the base of Q1401 and exits at connector 1D4 to the stereo decoder module.

Stereo Decoder Module

Refer to the Fig. 7-14 as you read the following section.

The composite audio input signal, which consists of the stereo and second audio program (SAP), is applied to the base of emitter-follower transistor Q1625. Signals which appear at the emitter of Q1625 are coupled to IC1601 through R1601 and C1602. IC1601 is the same type of stereo decoder used in standard FM receivers with some slight modifications. The output from pins 4 and 5 of IC1601 contain L+R and L−R signals. The output at pin 7 drives a stereo LED indicator when the pilot stereo signal is detected. A1604 is a 78.67 kHz trap which attenuates the second audio and reduces crosstalk between the stereo and second audio.

R1611 adjusts the free-running frequency of the phase-locked-loop (PLL) contained in IC1601. IC1601 also contains a stereo switch and a decoder which provides L+R and L−R at its output. The L+R and L−R signals are passed through low-pass filters A1602 and A1603. The L+R signal is routed through matrix decoder R1619, R1620, R1603, and R1607 and amplified by IC1609. The L−R signal is passed onto the dbx decoder. R1618 is a L−R level adjustment to correct for gain differences in stereo decoder IC1601. The L−R signal then passes through Q1617 and Q1618, which are electronic switches, and allow either second audio programs or the L−R audio to go to dbx decoder IC1607. R1703 adjusts the timing of the dbx circuit and R1709 is a highband trim adjustment.

The purpose of the dbx decoder is to restore the audio to its original characteristics (noise reduction). In transmitting, the audio is compressed or encoded, and then expanded or decoded at the receiver. This process is accomplished to enhance the signal-to-noise performance. From the dbx decoder at pin 15, the signal is passed on to dual-amplifier U1608 at pin 2 via R1769 and C1720. The signal is amplified by half of IC1608 and then returned to dbx decoder pin 5 through C1707 and R1704. The L−R signal exits at pin 8 through C1618 to the second half of U1608 at pin 5, where it is amplified and exits at pin 7. The amplified signal is applied to matrix amplifier IC1609 together with the L+R audio signal from C1610 and R1603. In the matrix amplifier and its associated circuits, the L+R and L−R signals are added and subtracted to give the L and R audio signals. R1626 and R1627 are the left and right matrix controls.

The left and right audio channels then go to four-channel stereo switch IC1603 at pins 13 and 11. The external audio channels are also applied to pins 2 and 4 of the stereo switch. When the SECOND AUDIO PROGRAM signal is selected, it is switched by Q1618 and applied to IC1603 pins 11 and 13 instead of the left and right audio channels. Audio inputs to U1603 are controlled either by dc voltages applied to connector 4B6, or electronically via Q1616 or Q1618.

Second Audio

The SAP at the emitter of Q1625 is applied to 78 kHz bandpass filter A1601. After A1601 filters the signal, it is amplified by Q1606 and Q1608 and routed to the carrier detector, CR1601 and CR1602. It is then rectified and applied to Q1605 to drive the SAP mute transistor Q1604 and the SAP presence indicator at pin 3 of connector 4Q6. SAP decoder IC1602 is a phase-locked-loop IC. Transformer T1652 and R1653 control the frequency of its internal oscillator. The 78 kHz FM signal is fed to pin 2 of the decoder. The demodulated output signal at pin 7 is filtered and amplified by Q1607 and Q1616 and the output level is set by R1671. The signal passes on to a dual emitter-follower dbx input switch at Q1617 and Q1618, which controls the dbx input, or L – R signal. When the SAP signal is applied to the four-channel stereo switch, the stereo inputs are shorted together internally.

Stereo Audio Processing

The fixed-level output of IC1603 is passed to connector 4B4 via R1749, C1679, R1750, and C1680. These same stereo audio channels are also fed to audio processor IC1604 where the bass, treble, volume, balance, and stereo enhancement are all controlled by dc voltages applied via connectors 4D6, 4A6, and 4B7. Transistors Q1610, Q1611, and Q1612 are used to match the dc volume controls of different Zenith systems to this processor. An improved stereo effect is accomplished by using a small amount of phase-reversed internal cross-coupling. This feature is activated by the EXTEND STEREO switch on the control panel via connector 4D6. Finally, the variable stereo audio at connector 4G3 drives the stereo power amplifier.

Chroma Circuit

Refer to Figs. 7-15 and 7-12 as you read the following sections.

The chroma signal is applied to the overdrive limit amplifier input at pin 10 of IC2376. This limiter limits signals of 1 V p-p or larger to aid in speeding the chroma APC pull-in. The limiter also provides a dc signal to the ACC pickup-noise canceller and the ac chroma signal to the ACC amplifier.

The ACC amplifier is a dc gain-controlled amplifier. In addition to the chroma signal, the ACC amplifier receives an ac signal for pickup-noise canceling and the dc gain-control signal from the ACC phase detector via the offset control stage.

The ACC amplifier provides two signal outputs. One is the ac gain-controlled chroma signal, and the other a dc signal to the color killer.

The chroma signal is capacitively coupled through C2378 from pin 12 to pin 14 on the IC. From pin 14, the signal is fed to an overload limiter, which limits very high-level signals to prevent the ACC and APC phase detectors from saturating.

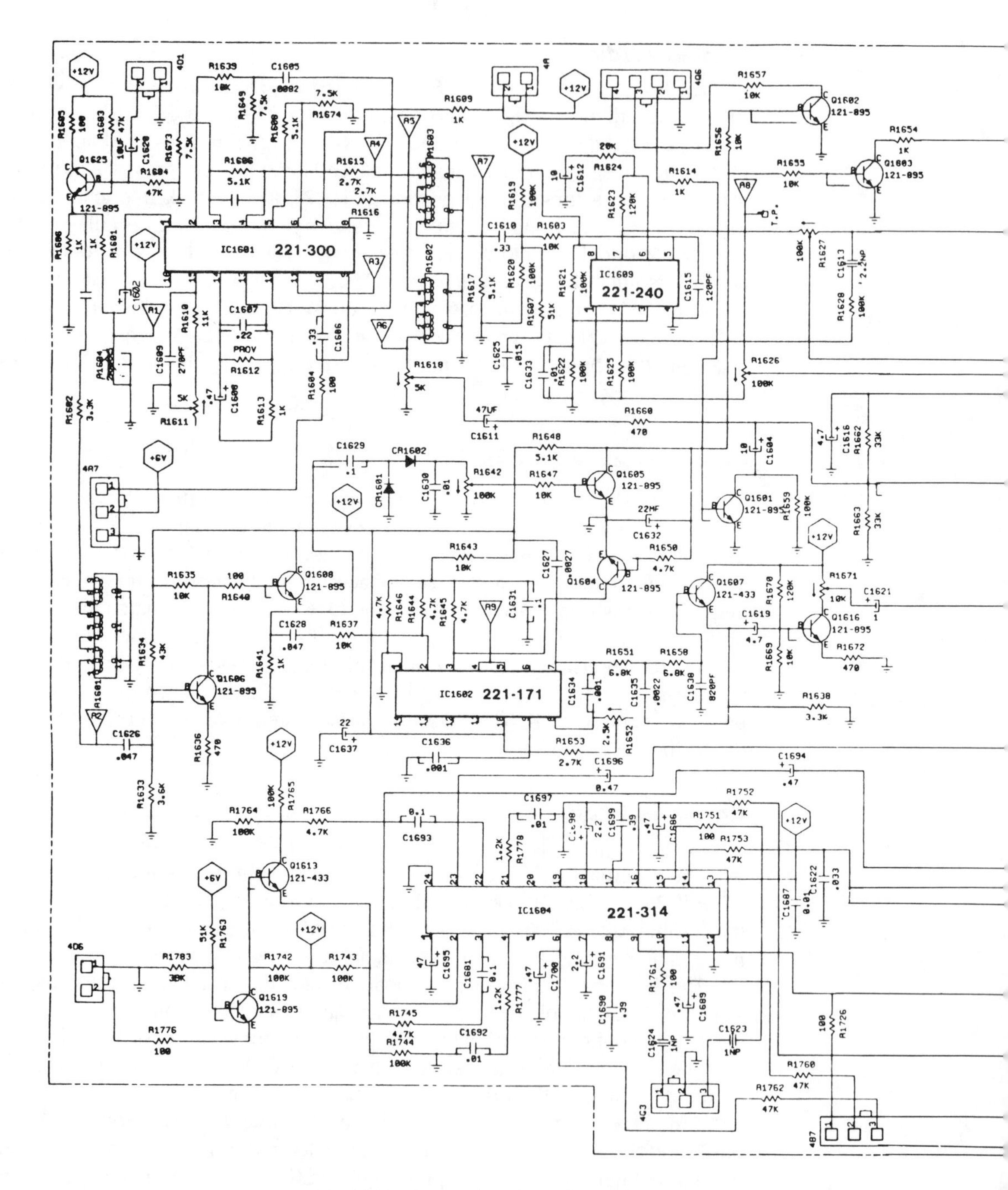

Fig. 7-14. Stereo decoder schematic.

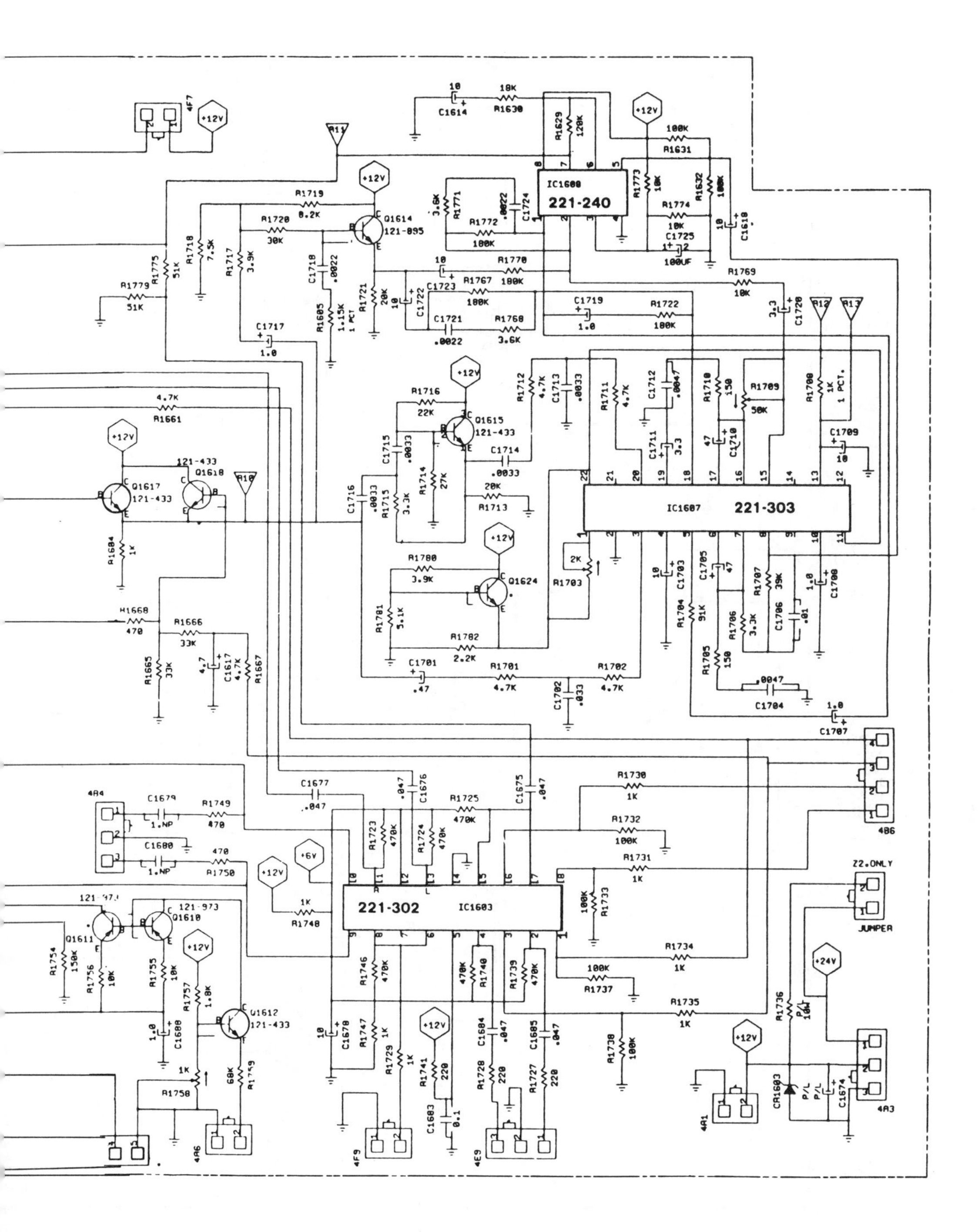

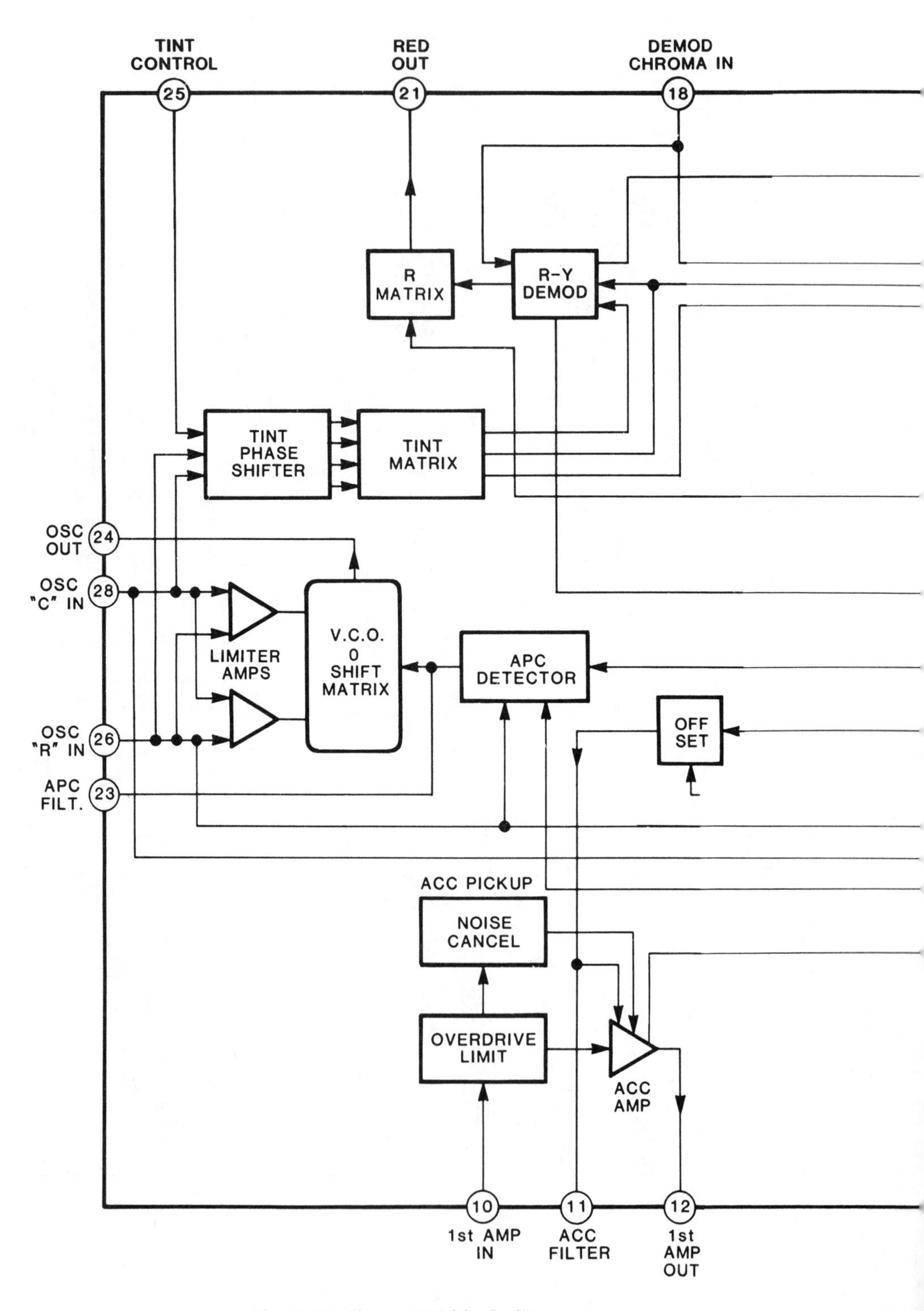

Fig. 7-15. Chroma IC block diagram.

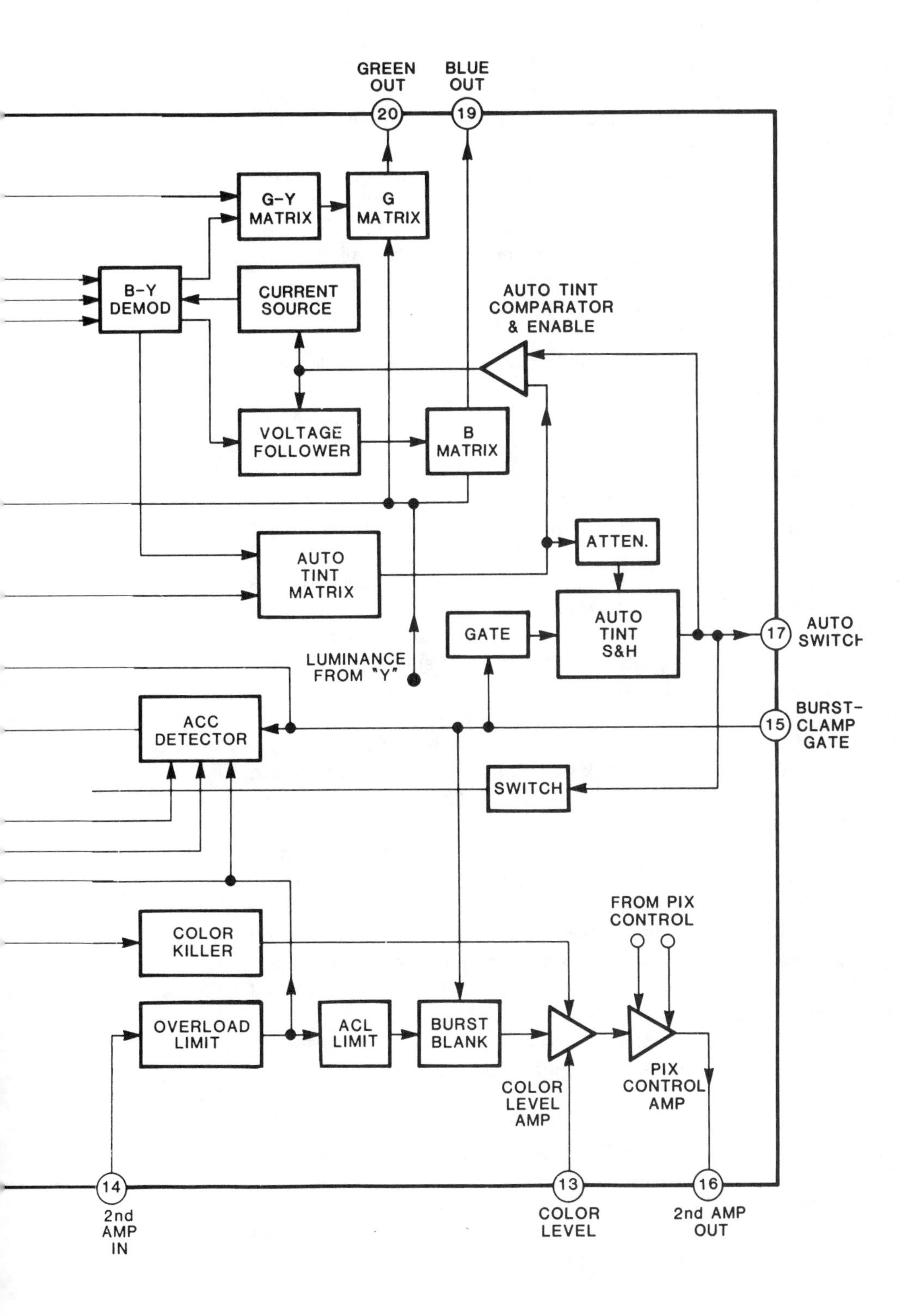
GREEN OUT
BLUE OUT
20
19
G-Y MATRIX
G MATRIX
B-Y DEMOD
CURRENT SOURCE
AUTO TINT COMPARATOR & ENABLE
VOLTAGE FOLLOWER
B MATRIX
ATTEN.
AUTO TINT MATRIX
GATE
AUTO TINT S&H
17
AUTO SWITCH
LUMINANCE FROM "Y"
ACC DETECTOR
15
BURST-CLAMP GATE
SWITCH
FROM PIX CONTROL
COLOR KILLER
OVERLOAD LIMIT
ACL LIMIT
BURST BLANK
COLOR LEVEL AMP
PIX CONTROL AMP
14
2nd AMP IN
13
COLOR LEVEL
16
2nd AMP OUT

From the limiter, the signal splits and goes to the automatic color level (ACL) limiter and the APC and ACC phase detectors. From the ACL limiter, the signal goes to the burst blanker. The blanker is driven by the burst/clamp gate, which deletes the burst from the signal to allow for the operation of the auto-tint circuitry. The chroma, without the burst, is then fed to the color level amplifier along with a dc signal from pin 13 (color level control) and a dc signal from the color killer.

The color-level amplifier gain is controlled by the pin 13 voltage. The color-level amplifier output feeds the chroma picture control amplifier, which is controlled by the picture control in the luminance circuit. The output at pin 16 drives single-tuned, chroma bandpass filter L2378 which, in turn, drives the demodulators at pin 18 of IC2376.

The color-killer circuit in the IC has no external filter. It receives a dc signal from the ACC amplifier that is similar to ACC gain. When the gain is sufficiently high, indicating low or nonexistent chroma burst, the color killer cuts off the dc current supply to the color level amplifier.

The APC and ACC phase detectors are double-balanced with push-pull outputs. Both detectors receive the chroma signal from pin 14 after the overload limiter. The reference signals are received from the voltage-controlled oscillator (VCO).

The ACC detector is an amplitude detector and the APC detector is a quadrature phase detector. The ACC detector output is fed through a switchable two-level offset to the ACC amplifier. Switching the offset changes the reference output level of the ACC amplifier, thus allowing a switchable ACL function.

The offset is switched by the auto switch at pin 17. The APC detector drives the VCO phase-shift matrix. The output is a dc voltage whose output is proportional to the phase angle between the VCO phase detector input and phase detector burst input. The external feedback path contains an output integrator, consisting of R2356 and C2359, a 3.58 MHz crystal, CR2356, and an input integrator, made up of R2355 and C2355.

The required matrixing for the tint-control circuit in IC2376 is all performed internally. The signal at pins 26 and 28 are fed to the tint phase-shifter. Here, the signals are dc controlled by the voltage at pin 25. The outputs of the phase shifter are four quadrature currents whose phase, relative to burst, can be changed by ±45 degrees. These currents are then matrixed to provide the reference phases for the demodulators and are phased 100 degrees apart. The output of the tint matrix is direct-coupled to the demodulators.

Two double-balanced demodulators with a G-Y matrix are used in the IC. Chroma is applied at pin 18 and the subcarrier is applied internally from the tint matrix.

G-Y is derived by matrixing the appropriate currents from B-Y and R-Y. The outputs at pins 19, 20, and 21 are positive blue, green, and red respectively.

The auto-color system performs three functions. It controls first, the ACL circuitry, second, the auto-tint circuitry, and third, the color killer threshold. It controls the ACL by changing the offset at the output of the ACC detector. This changes the burst level, and correspondingly changes the chroma level to limit over-saturated colors in the auto mode. The change in offset voltage also changes the color-killer threshold, allowing the viewer to choose, by means of the AUTO switch, the level of color kill.

The auto-tint matrix combines portions of B-Y and R-Y to give an "activation" area from about 30 degrees. The matrix is fed to the auto-tint comparator and an attenuator. The attenuator establishes a voltage that is proportional to the operating voltages of R-Y and B-Y demodulators. The sample-and-hold establishes a voltage on a capacitor (C2384) at pin 17 which is the reference voltage for the comparator. The other voltage on the comparator is higher during the burst interval because it is not attenuated. This offset disables the enable circuit. If, during the scan, there is nothing in the 30-degree to 210-degree semicircle being demodulated, the comparator is disabled and the system does not correct.

If a color is present in the semicircle, the comparator enables the system and the voltage follower and current source are activated. The voltage follower reduces the gain of the B-Y demodulator. This action makes the R-Y axis the preference axis. Colors that are away from the R-Y axis during activation are pulled toward R-Y. The current source draws current from the B-Y output operating point. This causes the R-Y corrected signal to be pulled to a flesh-tone corrected signal.

Luminance Circuit

The input to the luminance circuitry at pin 2 of IC2376 is driven by a 1.5 V p-p video signal from delay line L2276. The input signal is internally buffered and drives a series RLC network that is used to derive the peaking and low-pass luminance signal.

Inductor L2277 of the RLC network is connected across pins 4 and 6 of the IC. Across this inductor, a second signal is produced, called the "f" signal. This "f" signal is a bidirectional pulse used as the preshoot/overshoot peaking signal. The low-pass luminance "f" signal appears across capacitor C2282 at pin 6. The internal circuitry of the IC can process the common mode luminance and differential mode peaking signals separately. This adds a variable amount of the peaking signal back to the luminance to control the picture sharpness.

The process of adding the peaking signal to the luminance signal is accomplished inside IC2376. Under minimum peaking conditions, pin 1 is at 12 V, and all the internal signal processing transistors conduct equally. In this state, the positive and negative peaking signals cancel each other and no peaking signal is present in the luminance signal at the video output. As the voltage at pin 1 is decreased, the additive process is decreased, and the amount of peaking signal present in the luminance at the video output increases.

The automatic brightness-limiter circuitry is also internal to IC2376 and controls the gain of the video circuitry. With no current flowing through the tertiary winding of the high-voltage transformer into the CRT anode, the voltage at pin 3 of the IC is normally at 3.66 V. This voltage is determined by a resistive divider network. As current starts to flow to the anode of the CRT, the current is drawn from the junction of resistors RX3210 and RX3206 on the sweep module. When the anode current reaches 1.5 mA, the voltage at pin 3 of the IC is normally 3 V, which activates the internal picture-control circuitry. This results in a reduction of the video gain and anode current.

Automatic CRT Tracking

Refer to Fig. 7-12 as you read the following section.

The automatic CRT tracking system automatically adjusts the cutoff of the three picture-tube guns. The functions of the circuit are in IC2501 on the main module. This integrated circuit can be referred to as the automatic tracking control. It also provides the proper gray scale and black and white tracking under all conditions.

Integrated circuit IC2501 consists of several sections. One color system will be described; the other two operate in the same manner.

Each channel consists of a block called the gate-controlled amplifier, which is an electronic circuit equivalent to the former background control. A dc voltage now controls the proper gray scale and drive levels. In addition to the gate-controlled amplifier, a sample-and-hold circuit has been incorporated. This circuit samples the CRT current at a certain predetermined time and compares the cathode current in the CRT to the internally generated reference. The sample-and-hold and reference signals sense any changes and automatically vary the bias and gain of the amplifier to maintain the correct gray scale.

Another block in the IC receives an input from the vertical amplifier and generates a sampling pulse. The purpose of this sampling pulse is to set the CRT at the approximate black level and to sample the current coming from the cathode. Also, it simultaneously turns on the sample-and-hold circuit, which checks the level of the CRT current against the reference and changes the sample-and-hold voltage and the output voltage accordingly.

The sampling pulse occurs at the end of the vertical blanking interval. By reducing the vertical height so that the raster is underscanned, you can observe the sampling pulse as a very thin gray line above the normal video at the top of the screen. The sampling pulse has a very short time interval. This interval is just before the onset of normal video, during which the IC is actually reading the cathode currents of the CRT and referring them to previous references, which are generated inside the IC. If a current is either too high or too low, a circuit capacitor charges or discharges accordingly. After the sample impulse is over, that capacitor holds a normal voltage and permits normal video to pass.

Three current references inside IC2501 are set to the proper values to produce the three color temperatures, as determined when the CRT is manufactured. Capacitors C2504, C2505, and C2506, which are connected to pins 1, 4, and 7, are the sample-and-hold capacitors. Pin 20 receives information from the vertical deflection circuit at the collector of transistor Q2148. Capacitor C2507 and series resistor R2515, as well as voltage-divider resistor R2504, are also connected to pin 20. These components serve to generate the internal sampling pulse from the trailing edge of the vertical retrace pulse to coincide with the approximate end of the vertical retrace period.

The voltage at pin 19 of IC2501 is normally low (near ground), but rises to about 6 V when the sampling pulse occurs, during the sampling period. The resultant pulse is coupled through diodes CR2505 and CR2506 over to pin 15 of IC2376. If the vertical retrace time becomes too long, the sampling pulse could be placed into the video region. If that should occur, the automatic CRT tracking system would not function because it relies on the blanking leveling from IC2376 to accomplish the automatic tracking. The connection of diodes CR2505 and CR2506 back to the blanking input ensures that whenever the automatic tracking sample pulse is present, the chroma image will be held in the blanking state. This ensures that the sampling occurs at a blanking level rather than during an actual video signal.

Lines 2, 5, and 8 of IC2501 are sampling input points. These inputs come from the collectors of the video output amplifiers Q2506, Q2504, and Q2505. The three CRT cathode currents are sampled by these collectors to the IC.

Vertical-sweep-related circuits are also incorporated on the main module. The vertical signal at pin 16 of IC2710 is connected through voltage divider R2391 and R2392 and across capacitor C2363. This RC network integrates the signal to derive a vertical pulse. This vertical pulse drives the base of transistor Q2302, which then goes into saturation during the vertical interval. The base of transistor Q2301 is then pulled down through R2394 and the stage conducts. This action produces a clean vertical interval signal to operate the automatic tracking.

The +12 Vdc voltage on the emitter of transistor Q2301 is applied through the transistor and its collector circuit to pin 16 of IC2376, the chroma amplifier emitter-follower output. The presence of +12 Vdc at this pin cuts off the chroma output during the vertical interval. Therefore, no chroma or noise can pass through the video system when the automatic CRT tracking is in operation.

Video Output Amplifiers

The video output amplifier relates to IC2501. Transistors Q2504 (green), Q2505 (red), and Q2506 (blue), are incorporated into the output circuitry. The output amplifier transistors are Q2503 (green), Q2502 (red), and Q2501 (blue).

The amplifier system is a common-emitter circuit whose transistors Q2501, Q2502, and Q2503 drive the CRT cathodes through transistors Q2504, Q2505, and Q2506. Adequate frequency response is obtained by the peaking components, capacitors C2501, C2502, and C2503, and inductors L2584, L2585, L2586, and L2587. These inductors are self-resonant at the second harmonic of the chroma, trapping out 7.2 MHz and preventing its appearance on the CRT screen. In addition, the inductors provide some video peaking in the 2 and 3 MHz range. The emitters of the output amplifier are connected together and are biased by 6.0-V zener diode CR2504.

All CRT cathode currents flow from connector 2C5 through transistors Q2504, Q2505, and Q2506. The current of the blue cathode flows from pin 6 of 2C5 through inductor L2586, resistor R2586, and transistor Q2506 to pin 2, which is a sampling input of IC2501 as previously described. Similar paths apply to the other two cathode currents.

On the positive transitions of the video signals, diodes CR2503, CR2502, and CR2501, located in the base-emitter circuits of transistors Q2504, Q2505, and Q2506, are turned on and conduct the video to the cathodes. Thus, the IC can determine the values of the CRT cathode currents without interfering with the passage of normal video coming from the transistors to the cathodes. Resistors R2510, R2511, and R2512, and capacitors C2522, C2521, and C2520 help suppress arcs that could damage the IC.

CRT Grid (G1) Voltage Generation

The fixed dc voltage for G1, the control grid of the picture tube, is developed in a special circuit that couples this voltage to pin 1 of connector 2C5. This circuit has two functions: to develop the G1 biasing voltage, and to provide spot-burn protection. The 200-V source is divided by resistors R2516 and R2517, while capacitor C2509 filters the voltage appearing at pin 1 of connector 2C5.

The other components of G1 voltage generator circuit form the spot-burn protection. Capacitor C2523 and resistor R2520 form a time constant of several seconds to maintain the G1 voltage after the receiver is turned off. The result is a very brief high-brightness condition that discharges the CRT anode voltage quickly when the set is turned off. This avoids the longer duration high-intensity spot on the screen.

Multivibrator Start-Up System

Refer to Fig. 7-16 as you read the following section.

The multivibrator start-up system "kick-starts" the horizontal output transistor and sweep circuit directly off the unregulated 150-V supply.

Transistors Q3213 and Q3224, together with capacitors C3213 and C3224, and resistors R3213, R3214, and R3215, form a stable multivibrator circuit operating off the 150-V supply. This voltage is developed on the main module when the power is turned on. Thus, the multivibrator also starts oscillating when the receiver is turned on.

The square-wave output voltage on the collector of transistor Q3224 varies between 150 V and ground. The frequency varies with ac line voltage, approximately 10 kHz to 20 kHz for an ac line variation of 80 to 120 Vac. The duty cycle of the square wave is approximately 70 percent on and 30 percent off.

The square wave output current through diode CR3216 and resistor R3216 is connected through connector 3X to an added winding on the horizontal driver transformer T3205 (located on the sweep module). The oscillating current through this winding is coupled to the driver transformer winding that is connected to the horizontal output transistor, and causes it to start conducting and to start the sweep circuit, which develops the operating voltages for continued operation of the set.

Once the receiver is operating, the start-up circuit is no longer needed and can be turned off. This turn-off is accomplished by the rest of the circuit shown. Involved are transistor Q3210, capacitor C3210, and resistors R3210 and R3211. Note that Q3210 is connected across the emitter and base of Q3224, one of the multivibrator transistors. When the receiver power switch is first turned ON, the 150 V is present at both ends of resistors R3210. Thus, both the emitter and the base of Q3210 are at the same potential and Q3210 does not conduct. As operation of the multivibrator continues, a charge begins to build up on capacitor C3210. As soon as this charge reaches the level of 0.7 V (Q3210 emitter more positive by this amount than the base), Q3210 will turn on and short out the emitter and base of Q3224, causing this transistor to turn off. The oscillation will also stop, as will the flow of current through the start-up winding of the driver transformer.

The multivibrator start-up system has then done its job and the rest of the receiver will take over, with normal drive to the horizontal output transistor and the development of the B+ operating voltages.

The time constant of C3210 and R3210 and R3211 determines the length of time that the multivibrator will oscillate. The usual time is about 130 milliseconds (ms).

Vertical Output Circuit

Refer to Fig. 7-12 as you read the following sections.

The sync separation and the vertical drive functions are done in IC2710. A sawtooth output voltage is produced on capacitor C2118 at pin 11 by charging and discharging it through constant current supplies. The discharge current is varied by the HEIGHT control. This changes the peak-to-peak swing of the wave because it always recharges to the same level. Pin 13 of IC2710 is connected through 3G3 to a resistor/capacitor network, which constitutes a feedback circuit. The yoke current is sampled by resistors RX2146 and RX2147 and the signal is coupled through capacitors C2131 and C2152 in the sweep circuit to pin 13 through the feedback circuit.

During the first half of the trace, the output from pin 14 of IC2710 drives transistor Q2149. The collector of Q2149 is coupled through resistor RX2119 to the base of transistor Q2122. This transistor is turned on and controls the current flow from the supply on its collector through pin 5 of connector 3G3 to the yoke. The other side of the yoke incorporates capacitor C2148, which is being charged at this time. As the trace time proceeds, the output at pin 14 of the IC gradually increases, turning transistor Q2149 on further. This reduces the current flow through transistor Q2122 until it reaches zero. At this point, the yoke current has also dropped to zero.

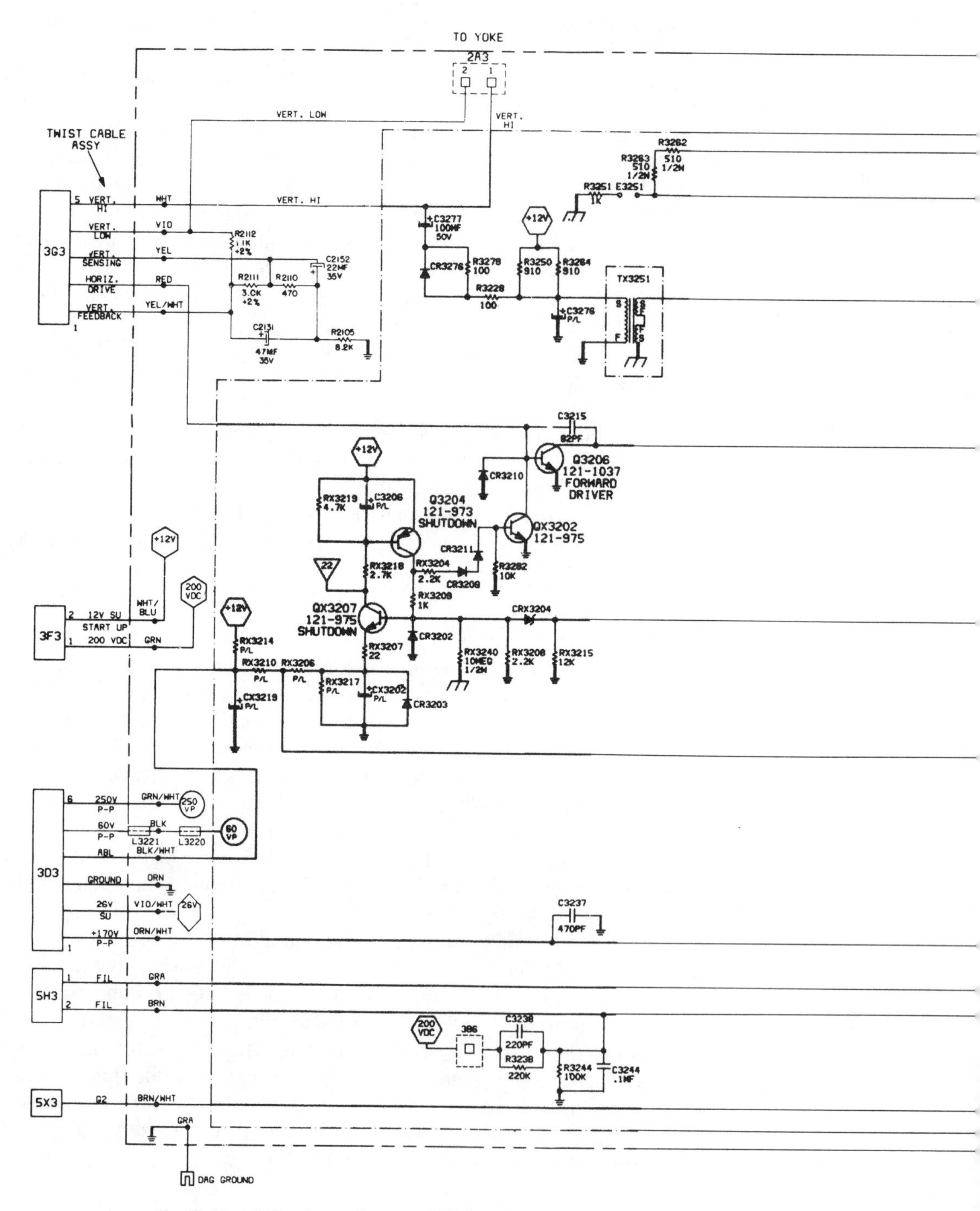

Fig. 7-16. HV/horizontal sweep schematic.

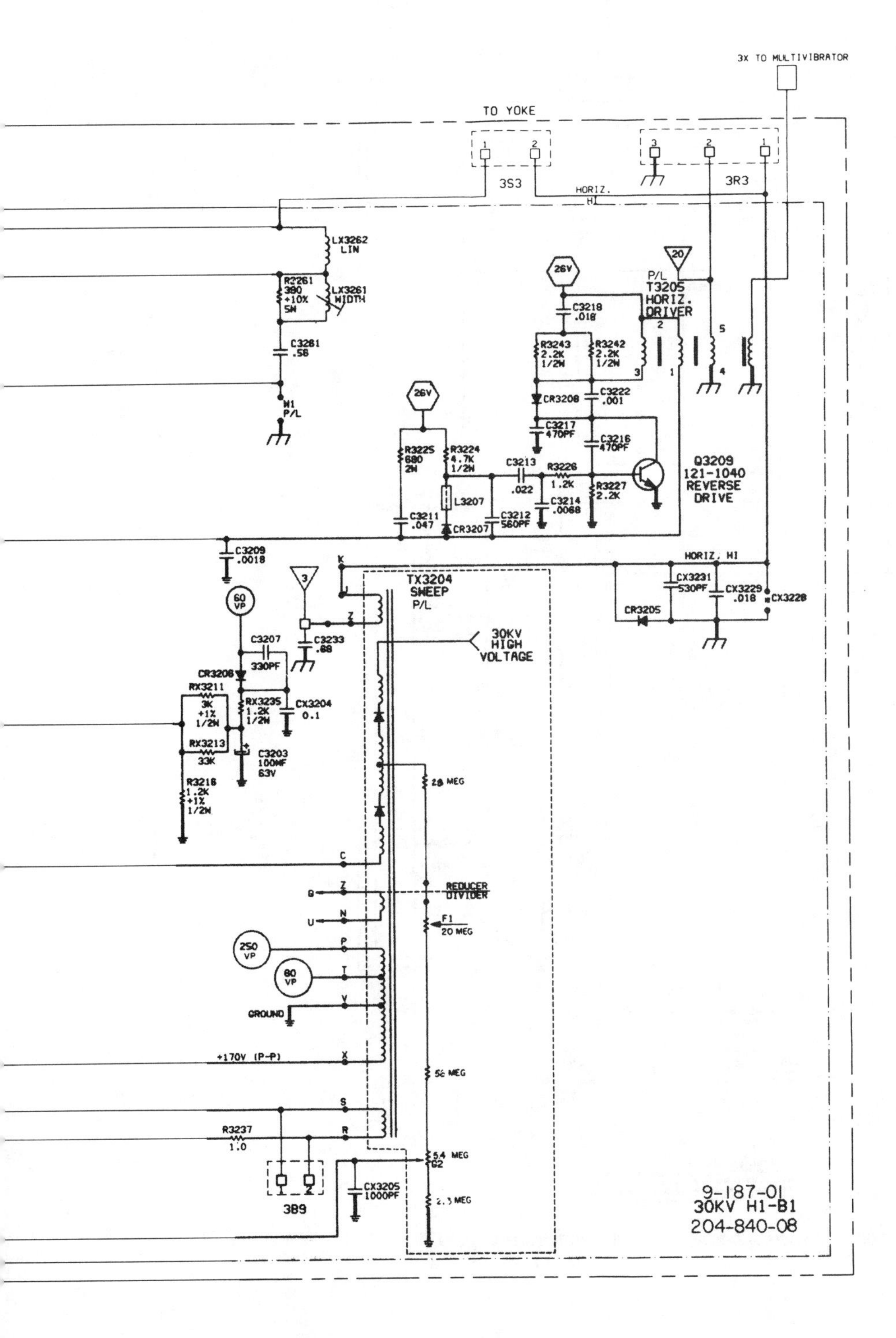
3X TO MULTIVIBRATOR
TO YOKE
3S3
3R3
HORIZ. HI
LX3262 LIN
R2261 390 ±10% 5W
LX3261 WIDTH
C3261 .56
W1 P/L
26V
C3218 .018
P/L T3205 HORIZ. DRIVER
R3243 2.2K 1/2W
R3242 2.2K 1/2W
CR3208
C3222 .001
C3217 470PF
C3216 470PF
R3225 680 2W
R3224 4.7K 1/2W
C3213 .022
R3226 1.2K
R3227 2.2K
C3214 .0068
Q3209 121-1040 REVERSE DRIVE
L3207
C3211 .047
CR3207
C3212 560PF
C3209 .0018
HORIZ. HI
CX3231 530PF
CX3229 .018
CX3228
CR3205
TX3204 SWEEP P/L
60 VP
C3233 .68
C3207 330PF
30KV HIGH VOLTAGE
CR3206
RX3211 3K ±1% 1/2W
RX3235 1.2K 1/2W
CX3204 0.1
RX3213 33K
C3203 100MF 63V
R3216 1.2K ±1% 1/2W
28 MEG
REDUCER DIVIDER
F1 20 MEG
250 VP
80 VP
GROUND
+170V (P-P)
56 MEG
R3237 1.0
5.4 MEG G2
CX3205 1000PF
2.3 MEG
3B9
9-187-01
30KV H1-B1
204-840-08

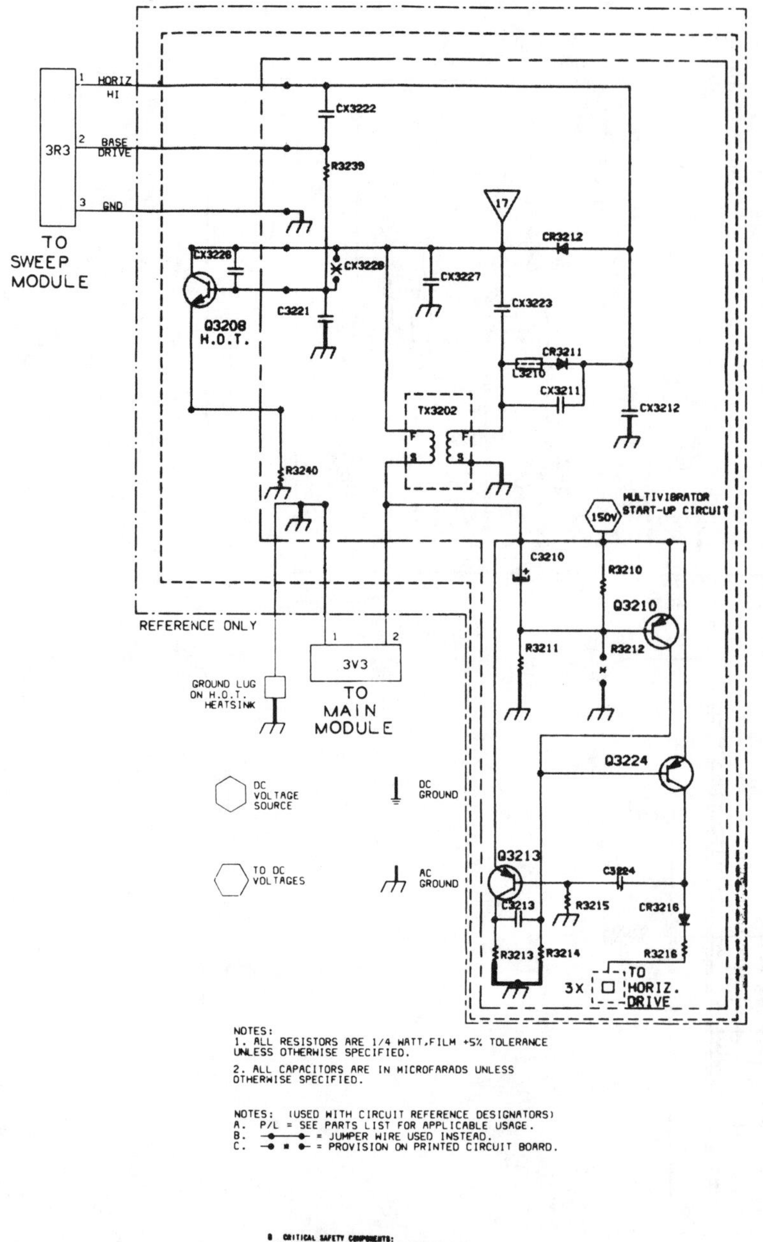

Fig. 7-16. HV/horizontal sweep schematic. Continued from page 133.

The second half of the trace is the time interval where the yoke current is going from zero to its peak negative value. The output of IC2710 pin 14 now drives Q2149 on even further. At this point, Q2122 is off and the current flows from capacitor C2148 (bottom of the yoke) through the yoke, through resistors R2122 and R2123, through diode CR2123, and through Q2149 to ground. Q2149 is gradually turned on further until maximum reverse yoke current is flowing at the end of the trace.

Transistor Q2147 has been off through all of the trace. Q2148 has also been off and capacitor C2149, connected to the collector, is charged to slightly over 25 V. When retrace starts, the output at IC2710 pin 14 goes low, turning Q2149 off. The yoke current does not stop immediately but flows through diode CR2119 into the power supply. This causes a sudden rise in the voltage at the connection to the yoke. The voltage increase is great enough to be coupled to transistor Q2147 and turn it on. Transistor Q2147 causes transistor Q2148 to turn on and go into saturation. The negative side of capacitor C2149 is, in effect, connected to the 26-V supply through transistor Q2148. Because the capacitor was fully charged, its positive terminal goes to approximately 50 V above ground. Diode CR2149 keeps the capacitor from discharging into the 26-V supply. The result is that the collector of the output transistor Q2122 is now operating from a 50-V supply.

With transistor Q2149 turned off, transistor Q2122 starts to conduct as soon as the yoke current reverses. When the yoke current reaches its proper value, the drive to Q2149 is turned on and the circuit is ready to begin the trace time again. At this point, transistors Q2147 and Q2148 turn off and C2149 starts to discharge. The circuit is now ready to start another trace period.

Vertical Countdown System

Refer to Fig. 7-17 as you read this section.

The vertical countdown system in IC2710 incorporates a G-counter, a clock phase-select circuit, and a reset and VM-generation circuit. The G-counter is set at state 380 and generates a 380 μs pulse, which is synchronous with the vertical sync pulse of off-the-air signals.

Other than during the vertical sync pulse period, the G-counter is reset at state 52 and starts counting at the leading edge of the next flyback pulse. The clock phase-select circuit compares the coincidence of the leading edge of the flyback pulse with the 64 fH (fH represents horizontal frequency rate) clock, and selects the appropriate clock phase for the G-counter.

Vertical Integrator. The composite sync from the sync separator output at pin 28 is integrated by a dual time-constant integrator to remove the horizontal sync pulses and higher order harmonics of the vertical sync. This integrated vertical sync pulse is fed to a shaping circuit via pin 1. The reconstructed vertical sync pulse is used for coincidence detection of the 525th counted pulse of the vertical counter in the phasing circuit.

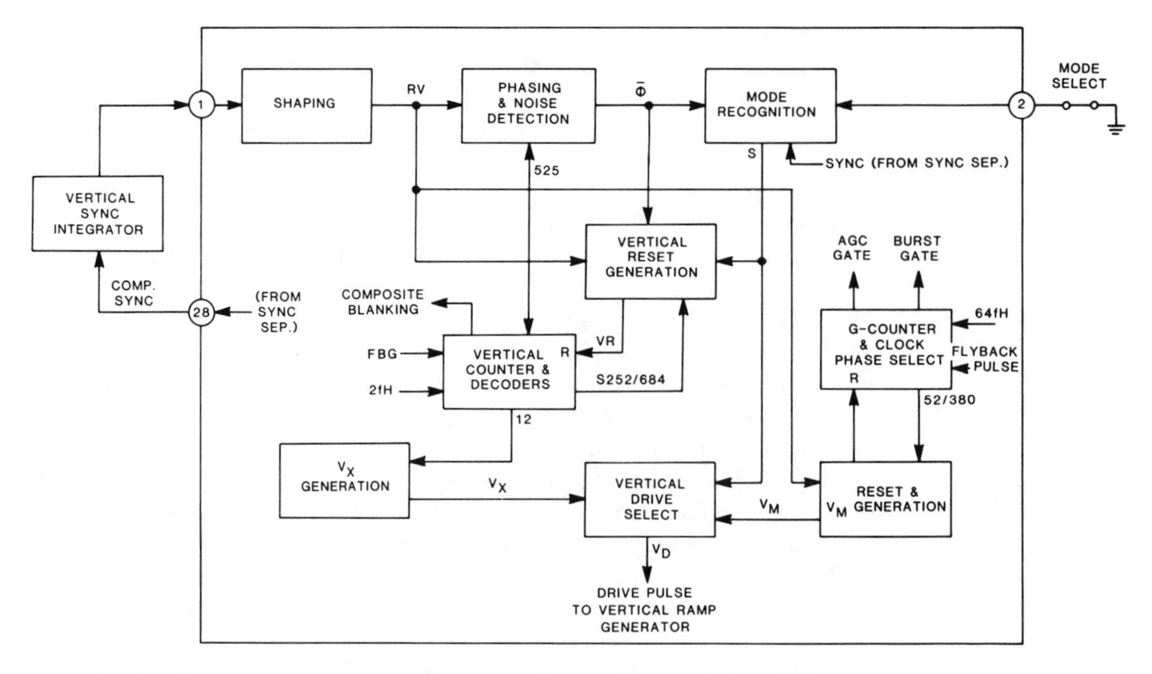

Fig. 7-17. Vertical-countdown circuit block diagram.

Standard Noisy Signal. A noise detector is enabled for a 184-line period following the vertical sync. Under a noisy condition, the detection of noise will force the system to operate in the standard mode to ensure good noise immunity.

Vertical Ramp Generator and Comparator

Refer to Fig. 7-18 while you read the following section.

The ramp generator receives its input from the vertical countdown section as a positive drive pulse (VD) measuring six horizontal lines wide (approximately 381 μs wide when in the standard mode of operation). When the pulse (VD) is low, it is buffered and isolated and prevents the ramp capacitor C2118 from discharging.

The level to which the ramp capacitor is charged is determined by a clamp circuit consisting of transistors and a zener diode. When the charge on the ramp capacitor reaches the clamp voltage level, the charging current is shunted to ground by the clamp circuit. As a result, the voltage is maintained at that level for the duration of the VD pulse.

The vertical height section consists of a current source that is controlled by the current flowing into pin 12 and the accompanying resistor/potentiometer network, which determines the current discharge of the ramp capacitor and the amplitude of the vertical ramp waveform.

The feedback from the vertical output is applied to pin 13 for linearity correction. Vertical drive is output at pin 14 to the vertical output section, described in Fig. 7-12.

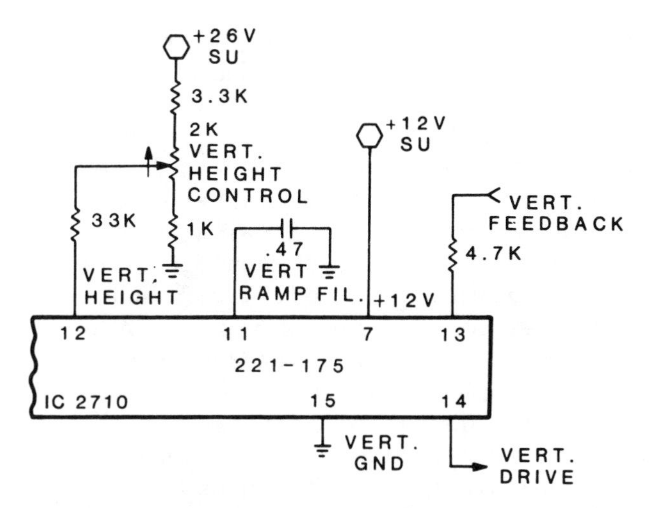

Fig. 7-18. Vertical ramp-generator/comparator IC block diagram.

Horizontal System

Refer to Fig. 7-19 as you read the following section.

The video signal is bandwidth-limited by a low-pass filter, consisting of R2732 and C2732, to attenuate frequencies higher than 400 kHz but not to severely degrade the transition time of the horizontal sync pulse.

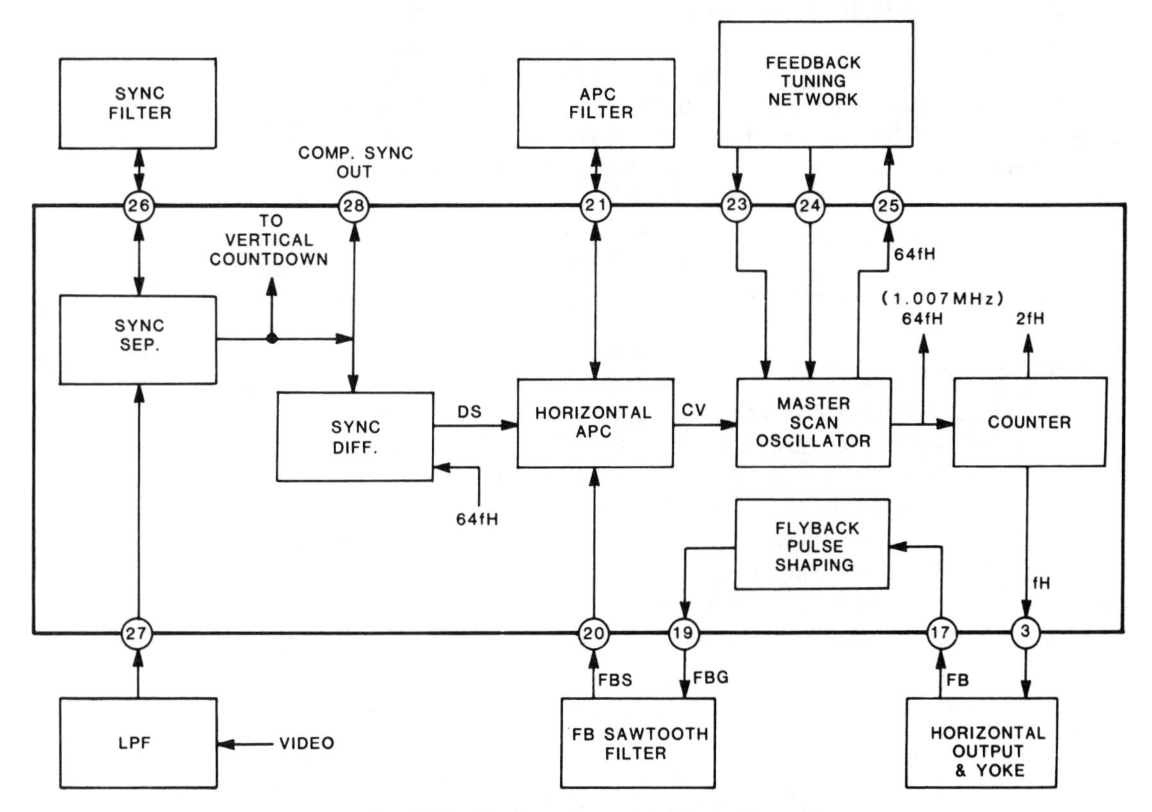

Fig. 7-19. Horizontal system block diagram.

Sync Separator, Filter and Differentiator. The sync separator has a variable slicing point that is selected to track the midpoint of the sync pulses in the video signal. The sync filter consists of a horizontal and a vertical filter.

Sync pulses are generated at the output of the sync separator, pin 28, while the sync filter is being charged. The broad pulses within the vertical sync pulse of the composite sync signal are differentiated to 6-μs pulses by the sync differentiator. This differentiated sync signal is used for gating the APC of the phase-locked-loop system. If there is no sync differentiation, the broad pulses during the vertical sync period would cause the PLL to make an incorrect adjustment, forcing the flyback sawtooth to be centered with the broad pulse, thus creating a frequency change at the VCO output. As a result, there may be errors in the AGC gating and burst gating of the vertical countdown system.

APC and APC Filter. The horizontal APC is basically a phase detector circuit. The differentiated sync signal (DS) is compared with the flyback sawtooth waveform FBS. If the SYNC is coincident with the sawtooth, an equal amount of charge/discharge current will flow in and out of the APC filter at any given horizontal sync period. As a result, constant control voltage is present at the output of the APC.

If the SYNC is not coincident with the sawtooth (a phase error for example), a lagging flyback sawtooth will cause more discharge current from the APC filter, resulting in a drop in the control voltage which controls the VCO to change in frequency and vice versa. The APC filter consists of resistors and capacitors connected to pin 21 of IC2710.

Master Scan Oscillator. The output of the VCO (master scan oscillator) has a free-running frequency of 1.007 MHz. The VCO has another output where the signal has the same frequency and is conditioned for interfacing. The frequency of this circuit is divided down to the horizontal rate (fH) by a divide-by-64 counter. Twice the horizontal frequency (31 kHz) is tapped from the counter for the vertical countdown system. The fH is buffered and fed to the horizontal deflection circuit via pin 3. The flyback pulse is fed to pin 17 and output to pin 19 via a flyback pulse-shaping circuit. The signal at pin 19 is further conditioned to a sawtooth waveform by a filter consisting of C2705, C2706 and R2705. The sawtooth is then fed to the horizontal APC via pin 20 to close the loop.

Horizontal Sweep System

Refer to Figs. 7-12 and 7-16 as you read the following description.

The bridge rectifier (CRX3271 through CRX3274), filter LX3224, and CX3249A and B provide 150 V to the primary of chopper transformer TX3202.

Horizontal output transistor Q3208, which is in series with the primary of TX3202, controls the stored energy by varying the on-time inversely to changes in B+. If the B+ decreases, the on-time must increase to maintain a constant energy level. For constant energy, current must be constant. The secondary winding of the transformer provides the step-up and phase inversion. During conduction of transistor Q3208, the polarity across the secondary is minus-to-plus. The negative voltage at the anode of load diode CR3211 prevents its conduction and energy is stored in the transformer.

After transistor Q3208 turns off, the voltage on the primary reverses. Therefore, the secondary voltage also reverses, forward-biasing load diode CR3211 and releasing the stored energy to the system.

The amplitude of the retrace pulse at the collector of Q3208, and also any voltage derived from this pulse on the flyback transformer secondary, are directly proportional to the energy supplied.

Pulse Width Modulator. A pulse-width modulated system provides regulation of the horizontal sweep and sweep-derived voltage for line and load variations.

The pulse-width modulator section of IC2710 operates in a closed loop wherein the +95-V supply is controlled by feedback from the flyback-derived +26-V power supply. The 26-V supply is essentially a measure of the 95-V supply, which in turn is essentially a measure of the high-voltage supply.

The pulse-width range of operation is determined by the limiter range set by the IC parameters and by the amplitude of the gating sawtooth waveform applied to pin 4. Maximum pulse width (PW) occurs toward low ac line voltage and the minimum pulse width occurs toward high ac line voltage.

The beam current feedback modifies the PW in such a way as to reduce the +95 V by about 3.5 V at maximum beam current. Otherwise, the +95-V supply would be maintained closer to the +95-V operating point, and thereby affect horizontal size tracking versus beam current. The high voltage to the picture tube is also regulated by the same sequence of events.

In-Phase Driver. This system is an energy storage circuit, storing energy during conduction and supplying output base drive during the off-time of Q3208.

In this system, the driver is part of an in-phase system. When forward driver Q3206 turns on, output transistor Q3208 turns on. The direct drive produces a constant drive level independent of the duty cycle.

A disadvantage is the negative drive current. This current is unacceptable for the fast fall time in the output transistor. To maintain a fast fall time in output transistor Q3208, reverse driver transistor Q3209 is added to the circuit. Forward driver transistor Q3206 is driven by IC2710, and reverse driver transistor Q3209 is driven by the collector voltage of Q3206.

Diode CR3207 and resistor R3224 form a clamp circuit that clips (or clamps) the peak of the collector voltage of Q3206. The waveform is coupled to the base of Q3209 by capacitor C3213 and resistor R3226. Transistor Q3209 turns on after Q3206 turns off. The current through Q3209, flowing through the primary of T3205, drives the base of output transistor Q3208 negative, producing a "clean" output pulse to switch Q3208 in about .5 μs.

After the completion of retrace time, the pulse-width modulator is prepared to decide how soon the transistor will turn on to begin the energy storage process. When operating normally, the transistor turns on approximately 18 μs after retrace time and continues until the end of the trace. The turn-on time will vary from 5 μs to 30 μs after retrace. This time variation compensates for line-to-load variation.

Transistor Q3208 is on for chopper operation during the conduction period of damper diode CR3205. Diode CR3212 from the damper diode to the transistor, called a blocking diode, is reverse-biased until the damper diode is off. When the cathode of the damper diode goes positive, the blocking diode is forward-biased, clamping the yoke voltage through transistor Q3208. The current through the transistor is a composite of the chopper current and positive yoke current. The transistor turn-off starts the retrace time. Transistor Q3208, the damper diode, and the blocking diode are off. The voltage at the collector of Q3208 rapidly goes positive due to the inductive "kick" of the regulator transformer. The polarity reversal of the primary winding in the regulating transformer is also seen on the secondary. The positive voltage at the anode of load diode CR3211, called the injection diode, forward-biases the diode into conduction.

The retrace is produced by the resonant circuitry in the deflection circuit. The added feature is the injection current. Injection diode CR3211 is forward-biased because the anode voltage is increasing faster than the cathode voltage. The injection current aids in charging the retrace capacitor to a constant peak voltage level. The injection diode adds the power lost from the 150-V supply.

During the second half of the retrace time, the retrace capacitor has reached its peak voltage and starts to discharge back into the yoke and flyback transformer TX3204. The injection circuit continues to supply energy during retrace time.

The collector voltage has the conventional retrace pulse of about 800 V, p-p plus a variable time and amplitude step. The step time is narrow at low line-voltage and wide at high line-voltage. The amplitude is also equal to the voltage of the unregulated B+ supply, which is dependent on the line voltage.

Damper diode CR3205 acts in a conventional manner. The damper current starts at the end of the retrace time and ends in the center of scan. Injection diode CR3211 continues to conduct until the pulse width modulator restarts with a requirement for more power.

High Voltage. During the collapse of the magnetic field around TX3204 and the horizontal deflection yoke, a very high voltage is induced into the secondary windings of TX3204. This voltage is rectified to a very high dc potential (30 kV) through a series of transformer windings and diodes. These are inside the flyback transformer. This high voltage is coupled to the anode socket of the CRT. A part of this voltage is coupled to the FOCUS control. This voltage is then coupled to the focus grids of the CRT.

All of the power supplies in the receiver are derived from the flyback transformer. Although the dc voltages on this transformer can be measured easily, special instruments are required to measure the complex output voltages.

Power Amp

Refer to Fig. 7-20 as you read the following section.

The TV set incorporates a 2-watt-per-channel stereo audio system. A fixed or variable level of audio from the stereo decoder module can be selected at the back-panel stereo jack.

The stereo amplifier consists of +12- and +24-Vdc power supplies, voltage regulator IC 442, and two audio output ICs, ICX4441 and ICX4442.

The power is supplied to the amplifier module by two secondary windings on transformer TX2. One secondary winding supplied ac voltage to a full-wave bridge circuit for the +24 Vdc supply. The other secondary winding supplied ac voltage to IC442 for the regulated +12 Vdc supply.

The left and right audio signals are input to connector 3G4 and sent to pin 1 of each power amplifier IC, via C4404 and C4409. Resistors R4409, R4404, and their associated components provide the proper bias for the amplifier ICs. Resistors R4407, R4411, R4406, R4417, and R4410 control the gain of each stage. The output of each IC is coupled to connector 3N4 by C4403 and C4422. Speakers are connected at 3N4 via the SPEAKERS/HI-FI switch on the stereo jack accessory.

Remote Control Power Supply

Refer to Fig. 7-21 as you read the following section.

The remote control power supply module performs two functions: it switches power to the main chassis via relay KX9401, and it also provides a regulated +12 Vdc to the keyboard. The diode bridge, consisting of CR9402, CR9403, CR9404, and CR9405, supplies −12 and +12 Vdc for the stereo interface and tuner control modules. Q9402 regulates the 12 Vdc where Q9401 is a switch and ON/OFF relay driver transistor.

Ac Power Supply

Refer to Figs. 7-21 and 7-22 as you read the following section.

The neutral side of the line is connected through connector 9R3 pin 1, and 9S3 pin 2 to the primary of transformer T1. It also connects to one side of TX301. The hot side of the line is connected through connector 9R3 pin 2 on the remote control power supply module, and then through a 3⁄16-ampere (A) fuse (FX9401) and connector 9S3 pin 1 to the other end of the primary of T1. This side of the line also passes through the contacts of relay KX9401, connector 9R3 pin 3, and connector 3V3 pin 1 and 2 to the other end of the primary of T301 through 2-A fuse FX301.

Ac voltage is always applied to transformer T1 whenever the line cord is plugged in and provides the +5 and +12 V to the tuning and remote control circuits. These voltages must be kept on in order for the remote control receiver to operate.

Transformer TX301 is a one-to-one isolation transformer that supplies ac line voltage through connector 3R8 pins 1 and 3 to the main module to operate the set.

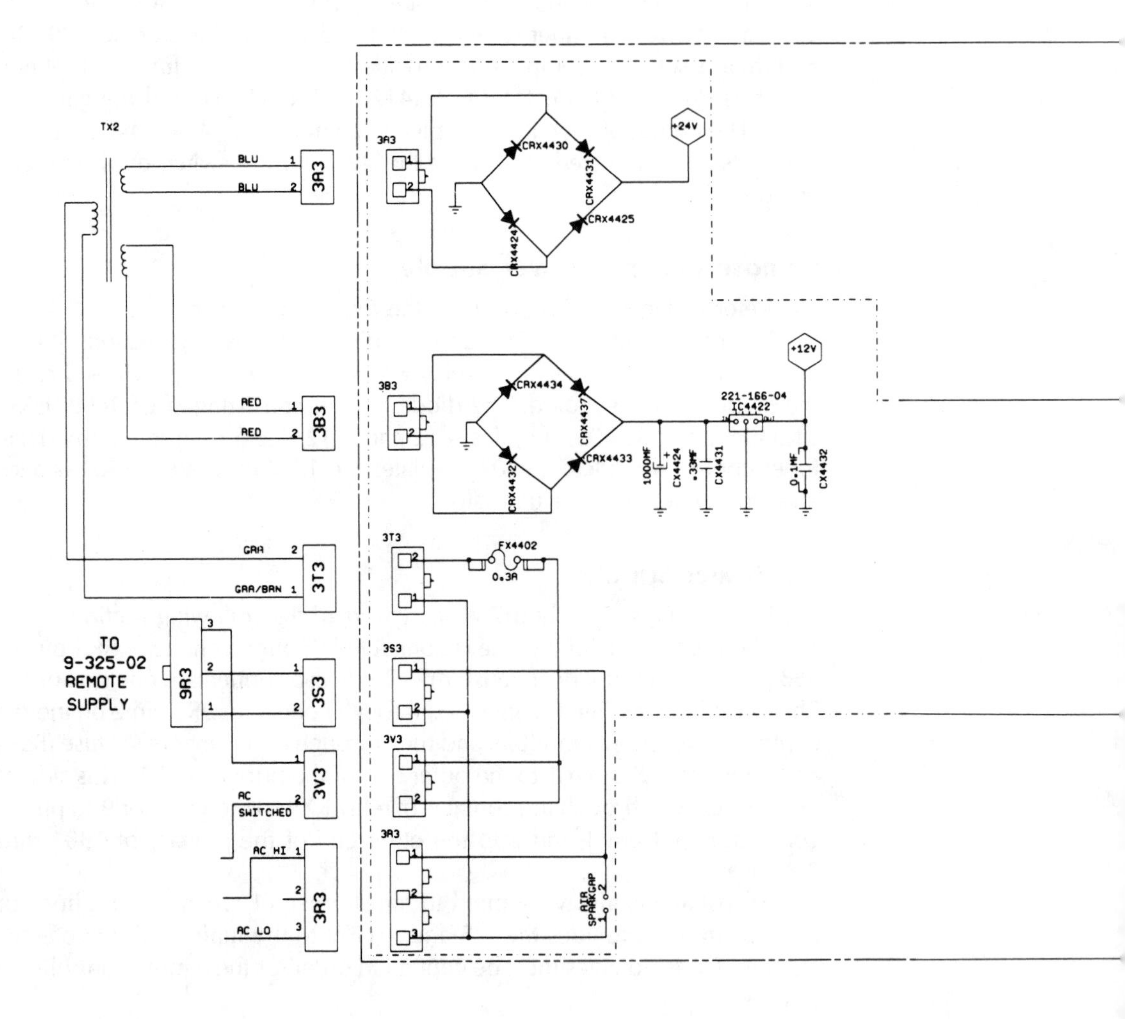

Fig. 7-20. Power-amp schematic.

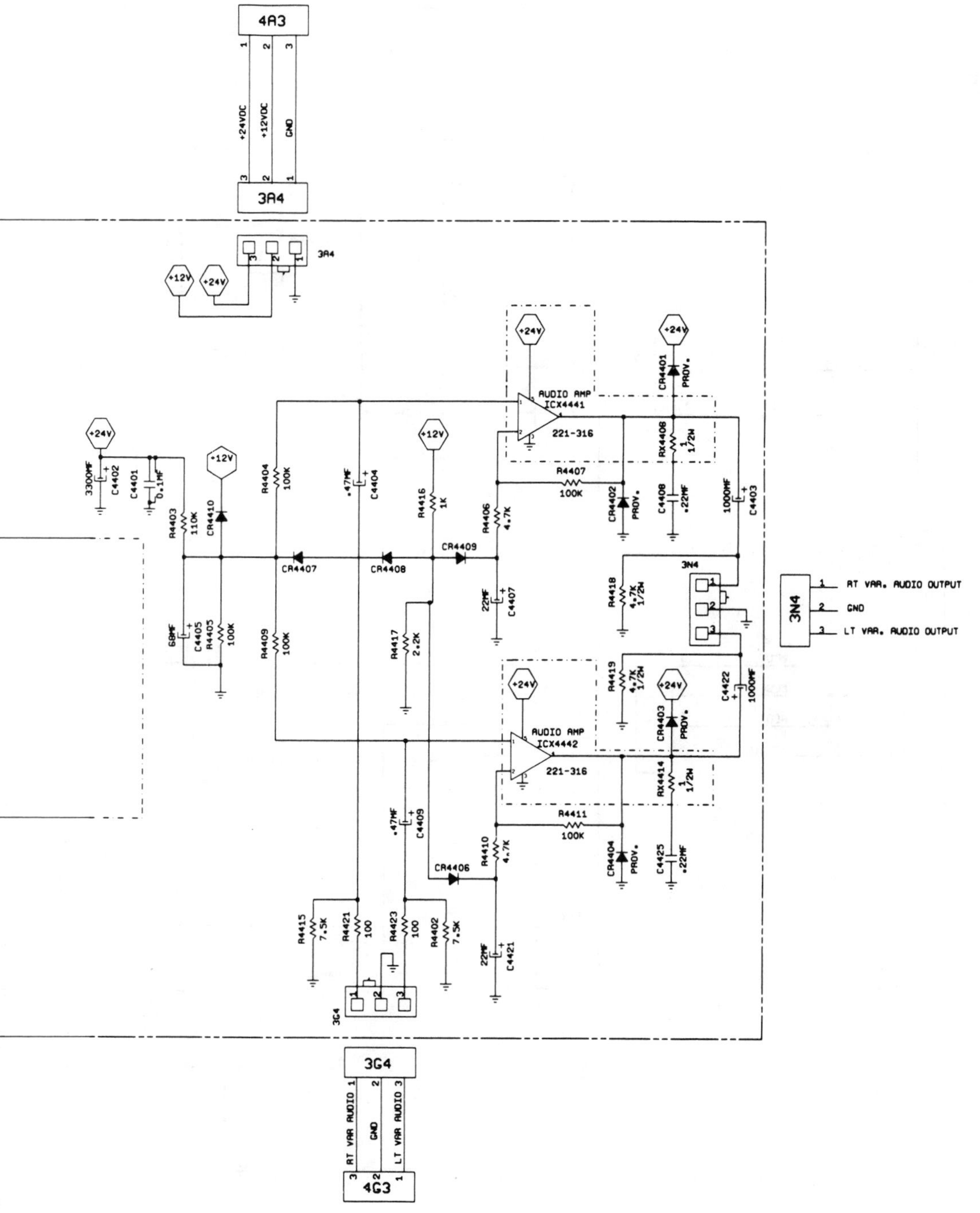
4A3
+24VDC
+12VDC
GND
3A4
AUDIO AMP
ICX4441
221-316
AUDIO AMP
ICX4442
221-316
3N4
RT VAR. AUDIO OUTPUT
GND
LT VAR. AUDIO OUTPUT
3G4
RT VAR AUDIO 1
GND
LT VAR AUDIO 3
4G3

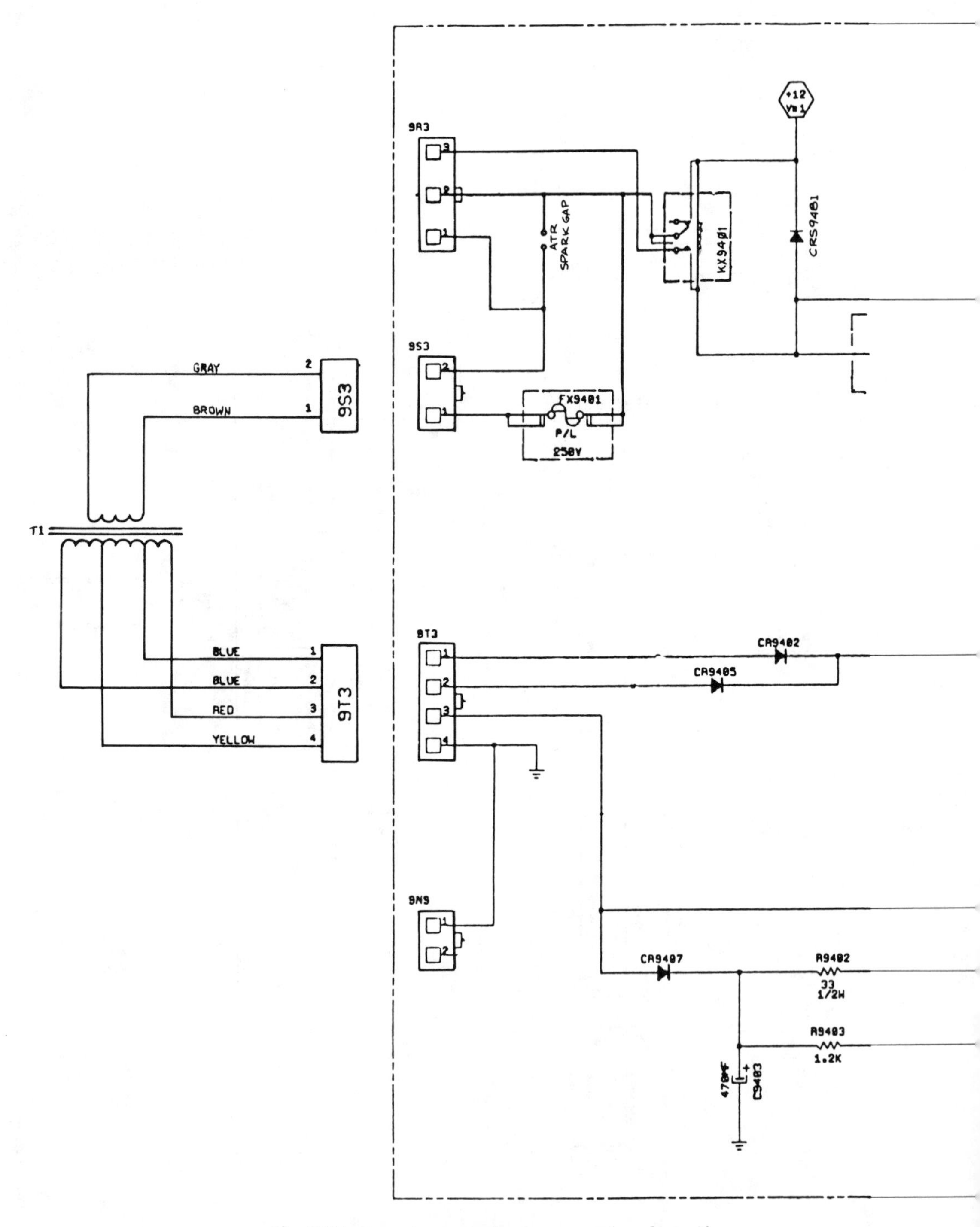

Fig. 7-21. Remote control power supply schematic.

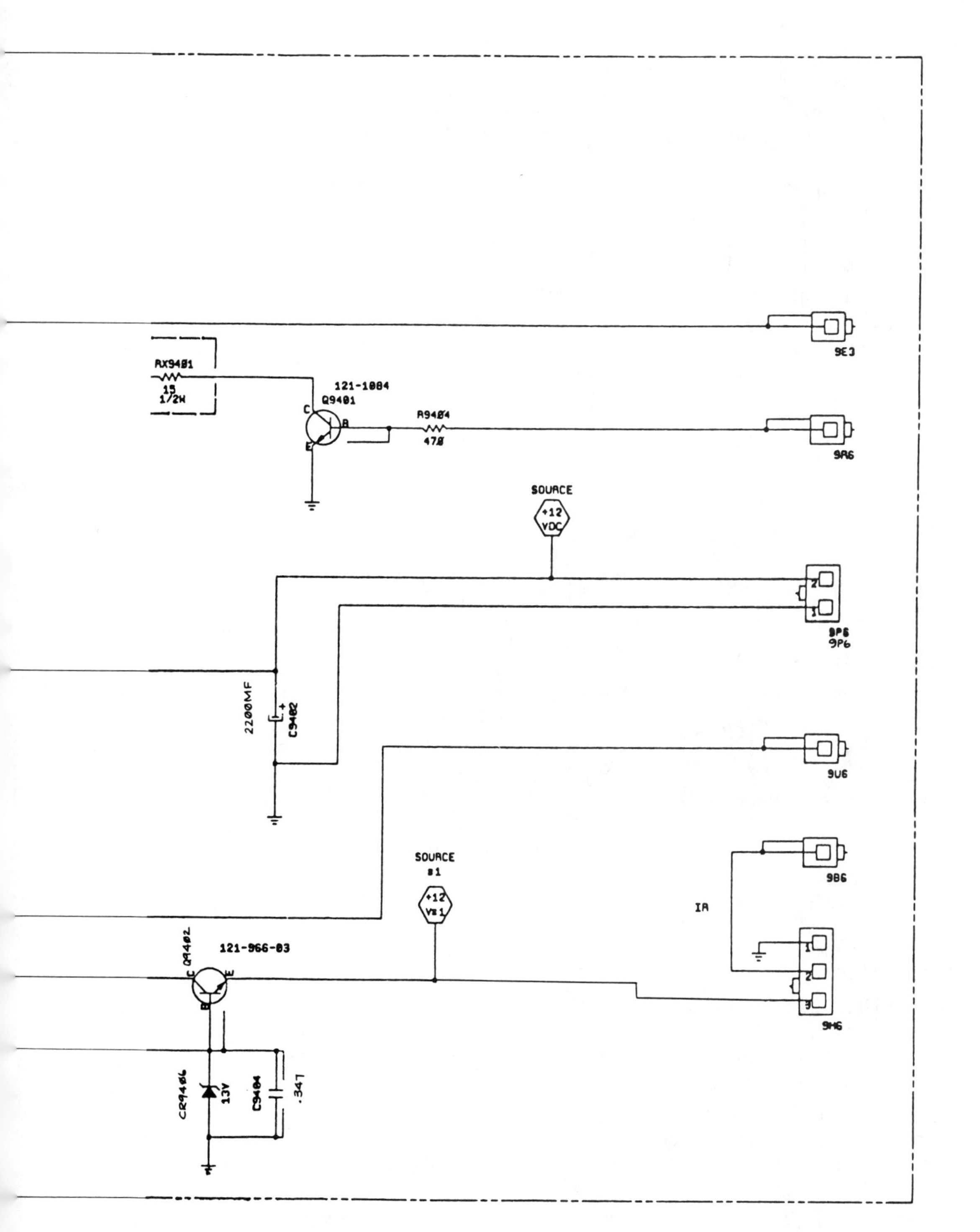

9E3
RX9401
15
1/2W
121-1004
Q9401
C
B
E
R9404
470
9A6
SOURCE
+12
VDC
2
1
9P6
9P6
2200MF
C9402
9U6
SOURCE
#1
+12
V#1
9B6
IA
Q9402
121-966-03
C
E
B
1
2
3
9M6
CR9406
13V
C9404
.347

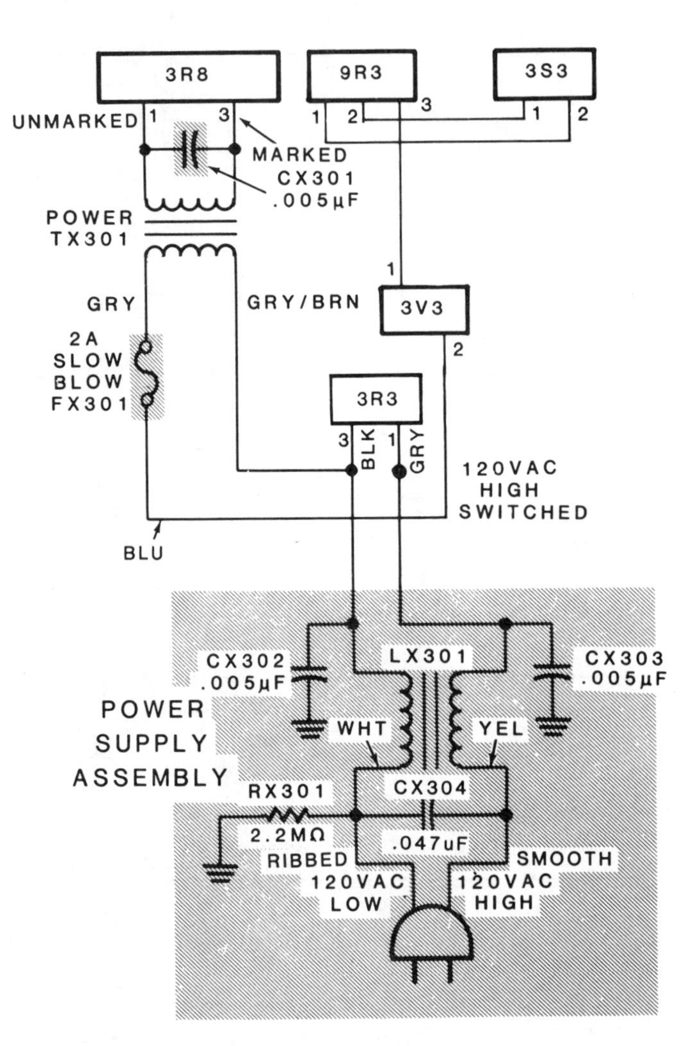

Fig. 7-22. Power supply schematic.

IR Transmitter

Refer to Fig. 7-23 as you read the following section.

Initially, the X ports (IC1 pins 1-5 and 16-18) are set to a logic-high output and the Y ports (pins 7-10) are set to sense the input state. When a key is pressed on the transmitter keyboard, a connection is made between the X and Y ports. The Y port senses the key closure and the oscillator is turned on, tuning for 37 milliseconds to allow for key debounce. Then the Y port senses the column in which the key is located. This information is stored in the X-Y decoder of IC1.

The X ports are then switched to the input state and the Y ports are placed in the output state at a logic-high level so that they can be sensed at the X port. The row in which the depressed key is located is sensed at the X port and stored in the X-Y decoder, and the key scan sequence is complete. At this point, The X-Y discriminator determines which key has been pressed and causes the appropriate command information to be transmitted to the TV set.

A "J" style, 6-V battery is used to power the transmitter.

Crosshatch Generator

Refer to Fig. 7-24 as you read the following section.

The crosshatch generator provides a crosshatch pattern for use during adjustments. The vertical signal from the main module at point C is shaped and amplified by transistor Q901 and IC U901D to provide square wave pulses during vertical retrace. Horizontal retrace pulses from R2704 on the main module are supplied to point E and are shaped and amplified by transistor Q904 and IC U901A to provide square-wave pulses during horizontal retrace.

Horizontal lines on the screen are generated by dual, up-counter IC U902. Vertical pulses at pins 7 and 15 reset the counters. After reset, horizontal pulses applied to pin 1 are counted. Every twentieth horizontal pulse causes pins 6 and 14 to become a logic high for one horizontal line time. Diode D901 and transistor Q905 are turned on, which applies the positive pulse to R2276 on the main module. This turns on the video output transistors and produces a white horizontal line on the screen. Because two vertical fields make up one vertical frame, a line is produced in each field and appears as a double line.

The vertical lines on the screen are generated by U903, which is connected as a resettable, astable multivibrator. Transistors Q902 and Q903 invert the vertical and horizontal pulses, respectively, which are then applied to U903 pins 3 and 13 to reset the multivibrator. Reset is also applied through resistor R918 to U901B pin 5. This reset operation causes the output of U903 pins 7 and 9 to be logic high. A very short time after, the reset is removed, which is determined by resistor R918 and the input capacitance of U901. Pin 5 of U901 becomes logic high and produces a logic low at IC U901 pin 4 and at the negative trigger input (pin 11) of one-half of U903. Pin 9 of U903 becomes logic low for approximately .4 μs (as determined by resistor R917 and capacitor C903). The logic low at pin 9 causes diode D902 to conduct, which produces a white vertical line on the screen in the same manner that the horizontal lines were produced.

When pin 9 returns to logic high, the rising edge of the waveform triggers the positive trigger input to the other half of the multivibrator. This causes U903 pin 7 to become logic low for approximately 4 μs (as determined by resistor R916 and capacitor C902). At the end of this period, when pin 7 goes back to logic high, U903 is again triggered at pin 11, and the process repeats. The vertical pulse applied to pins 3 and 13 of U903 prevent the multivibrator from operating during vertical retrace.

With the NORMAL/CROSSHATCH switch in the NORMAL position, 26 Vdc is an open circuit. In the CROSSHATCH position, the 26 Vdc from point F is supplied through resistor R921 to zener diode D905, which produces 16 Vdc for crosshatch operation.

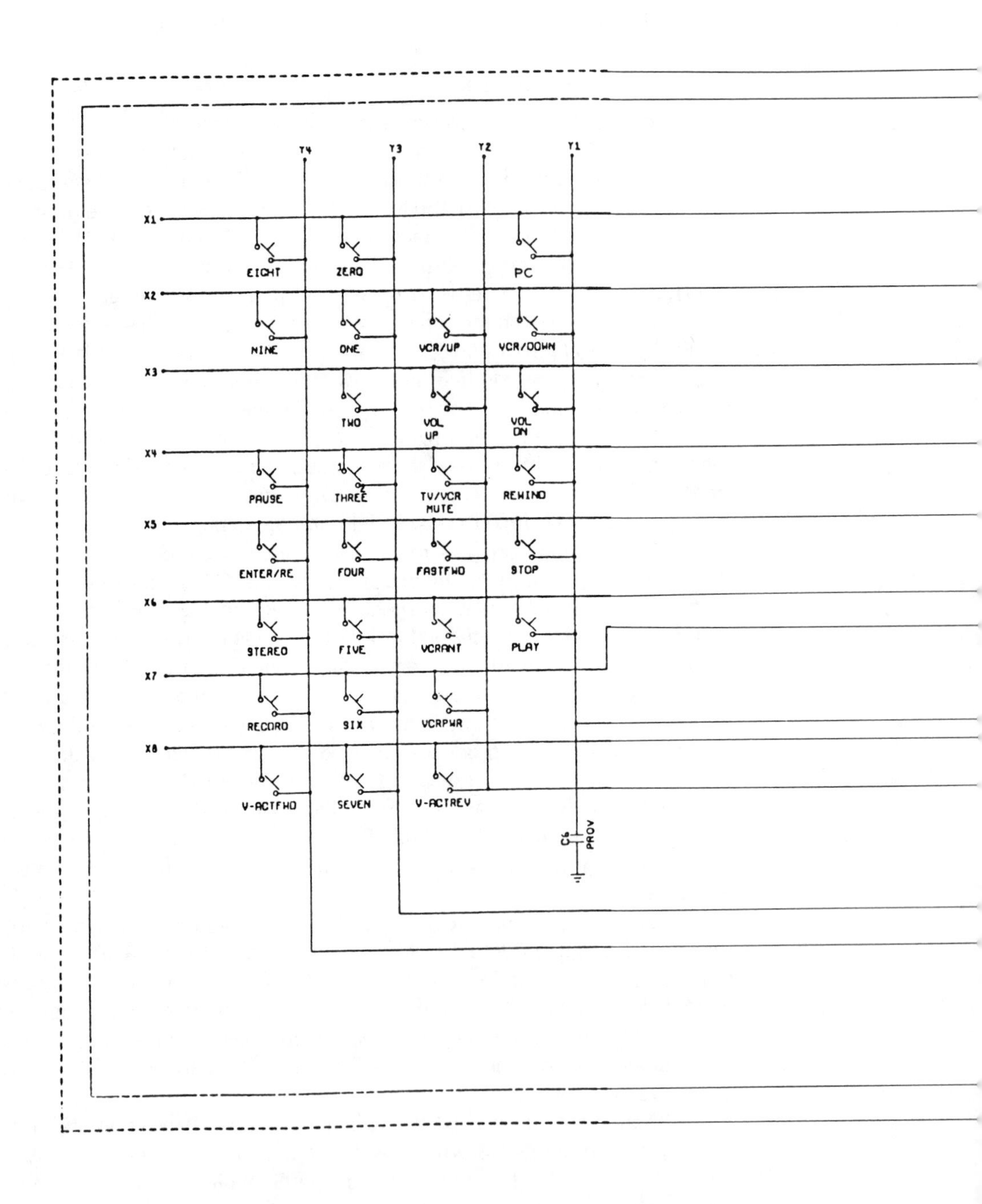

Fig. 7-23. Remote control transmitter schematic.

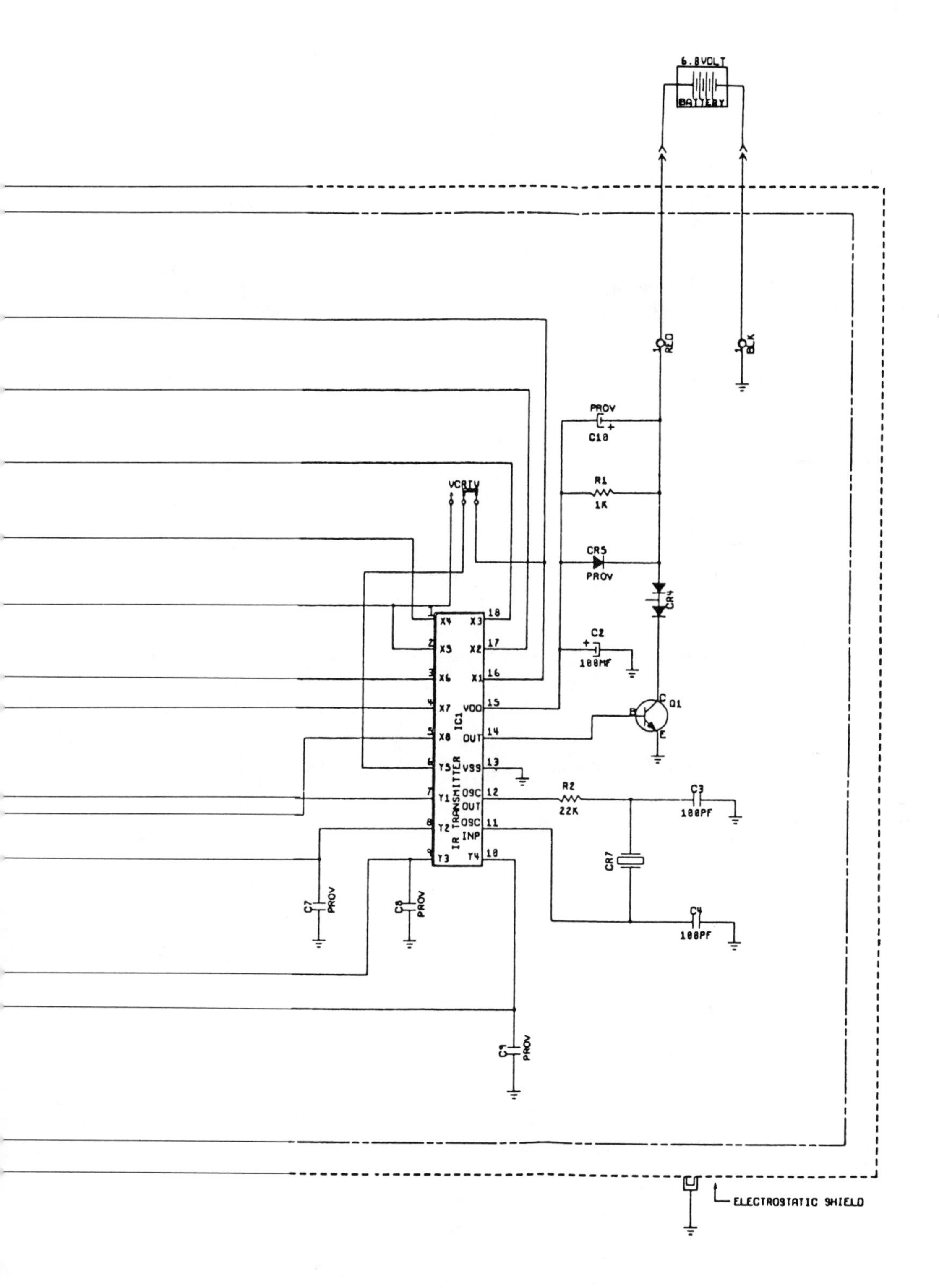
6.8VOLT
BATTERY
RED
BLK
PROV
C10
R1
1K
CR5
PROV
CR4
C2
100MF
Q1
VCR TV
IC1
IR TRANSMITTER
X4
X3
X5
X2
X6
X1
X7
VDD
X8
OUT
Y5
VSS
Y1
OSC OUT
Y2
OSC INP
Y3
Y4
R2
22K
C3
100PF
CR7
C4
100PF
C7
PROV
C8
PROV
C9
PROV
ELECTROSTATIC SHIELD

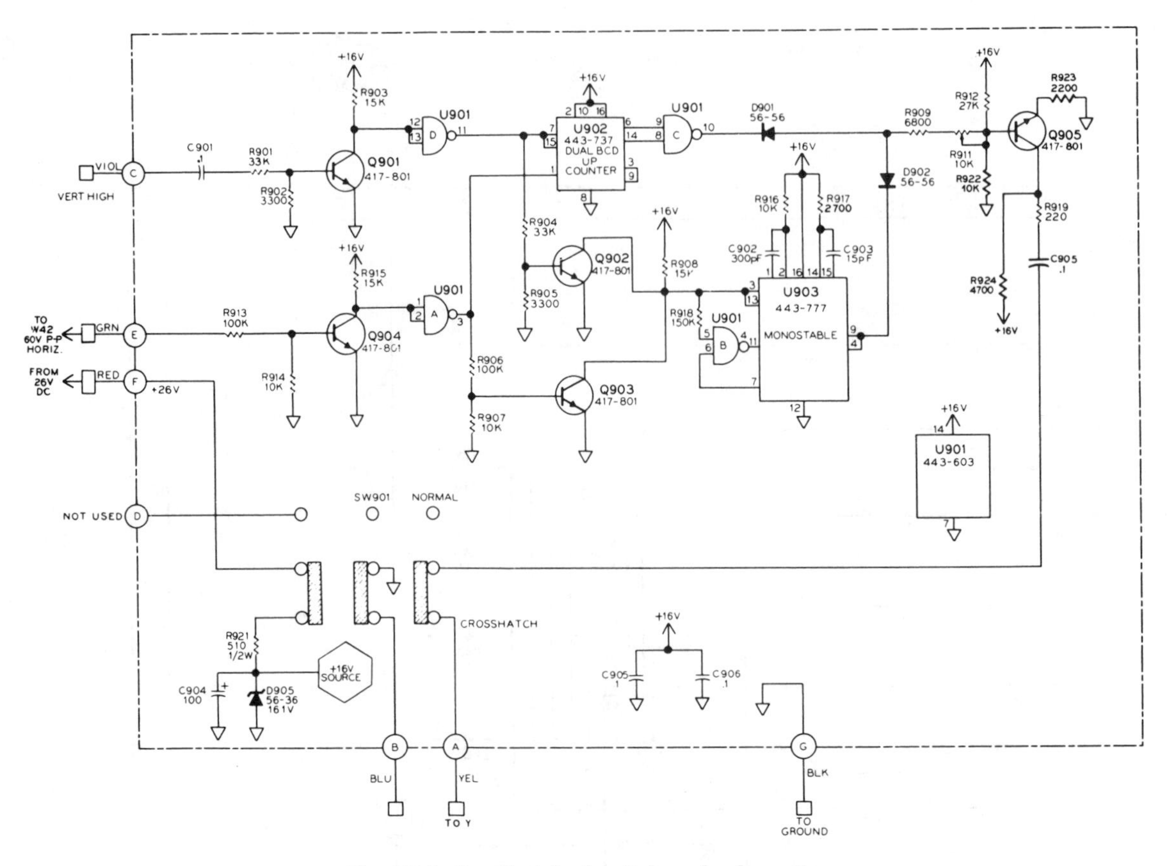

Fig. 7-24. Crosshatch circuit board schematic.

8
Specialized Electronic Equipment

One of the most useful pieces of electronic equipment is the oscilloscope. Oscilloscopes are used to troubleshoot and repair virtually every type of electronic device. The oscilloscope can accurately display a wide range of measurements that are encountered in electronics. Oscilloscopes are frequently used to display the amplitude and frequency of an electronic signal as well as the voltage level.

THE HEATHKIT DUAL-TRACE OSCILLOSCOPE, MODEL IO-4210 10-MHZ*

An oscilloscope is one of the first pieces of equipment that any technician or serious hobbyist wants to add to his or her collection of test instruments. There is no better way to learn about electronics and at the same time have the enjoyment of building your own oscilloscope, like the Heath Model IO-4210. The following information on this dual-trace oscilloscope was furnished courtesy of the Heath Company.

This oscilloscope is a portable, triggered-sweep, dual-trace dc-to-10 MHz laboratory-grade instrument. Outstanding features such as fast vertical rise time, good trace brightness, and high input sensitivity make the oscilloscope ideal for the wide range of measurements encountered in electronics, development laboratories, and scientific research. In addition, the rugged construction and dependable operation make it a versatile tool for either the hobbyist or the service technician.

Each of the two identical vertical input channels provides a maximum signal sensitivity of 10 millivolts/centimeter (mV/cm). Their attenuator networks can be switched through 11 calibrated ranges to set the deflection factor from 10 mV/cm to 20 V/cm.

You can select several modes of signal display by the VERTICAL CHANNEL MODE switch, POSITION control, and the TIME BASE switch. Either or both channels can be displayed as a function of time or as a function of each other. At lower sweep speeds, the vertical channel signals are alternately displayed at an approximate 200 kHz, chopped-mode rate, so both signals appear as a function of the same time base. For faster sweep speeds, both signals are displayed alternately on successive sweeps. During X-Y operation, the channel Y1(X) circuits provide horizontal X-axis deflection and the channel Y2(Y) circuits provide vertical Y-axis deflection.

Calibrated time-base ranges from .2 seconds/centimeter (s/cm) are readily switched in a 1-, 2-, 5-step sequence. The VARIABLE control provides variable sweep speeds between switch positions. Any sweep speed can be expanded five times when the X5 function is selected, providing a maximum sweep rate of 40 ns/cm.

The TRIGGER SELECT switch and LEVEL control allow the time base to be precisely triggered at any point along the positive or negative slope of the trigger signal. Various trigger signals can be selected. These include a sample of Channel Y1 or Channel Y2 input signals, an externally applied trigger signal, or a sample of the line voltage. The TRIGGER MODE switch controls the trigger input bandpass. A special TV position cuts off unwanted high-frequency signals. This is useful when you want to trigger on TV vertical frame signals.

A calibrated 1-V peak-to-peak square wave signal is provided through a front-panel connector, to allow easy probe compensation, vertical amplifier calibration, and comparison.

Front-panel display controls include INTENSITY, FOCUS, and VERTICAL and HORIZONTAL position. An additional control, accessible through the rear panel, allows you to adjust the ASTIGMATISM. A component tester allows you to check in-circuit or out-of-circuit electronic components.

Thus, this oscilloscope combines the most desirable features required for precise measurement and display, while its solid-state circuitry provides excellent sensitivity, stability, and versatility.

To get the maximum benefit from the information in this section, you should refer to the block diagram (Fig. 8-1) and also to the different sections of the schematic diagram (Fig. 8-2). The components are numbered in the following groups:

1- 99 Parts on the chassis.
100-199 Parts on the vertical circuit board.
200-299 Parts on the horizontal circuit board.
300-399 Parts on the low voltage circuit board.
400-499 Parts on the high voltage circuit board.
500-599 Additional parts on the horizontal circuit board.

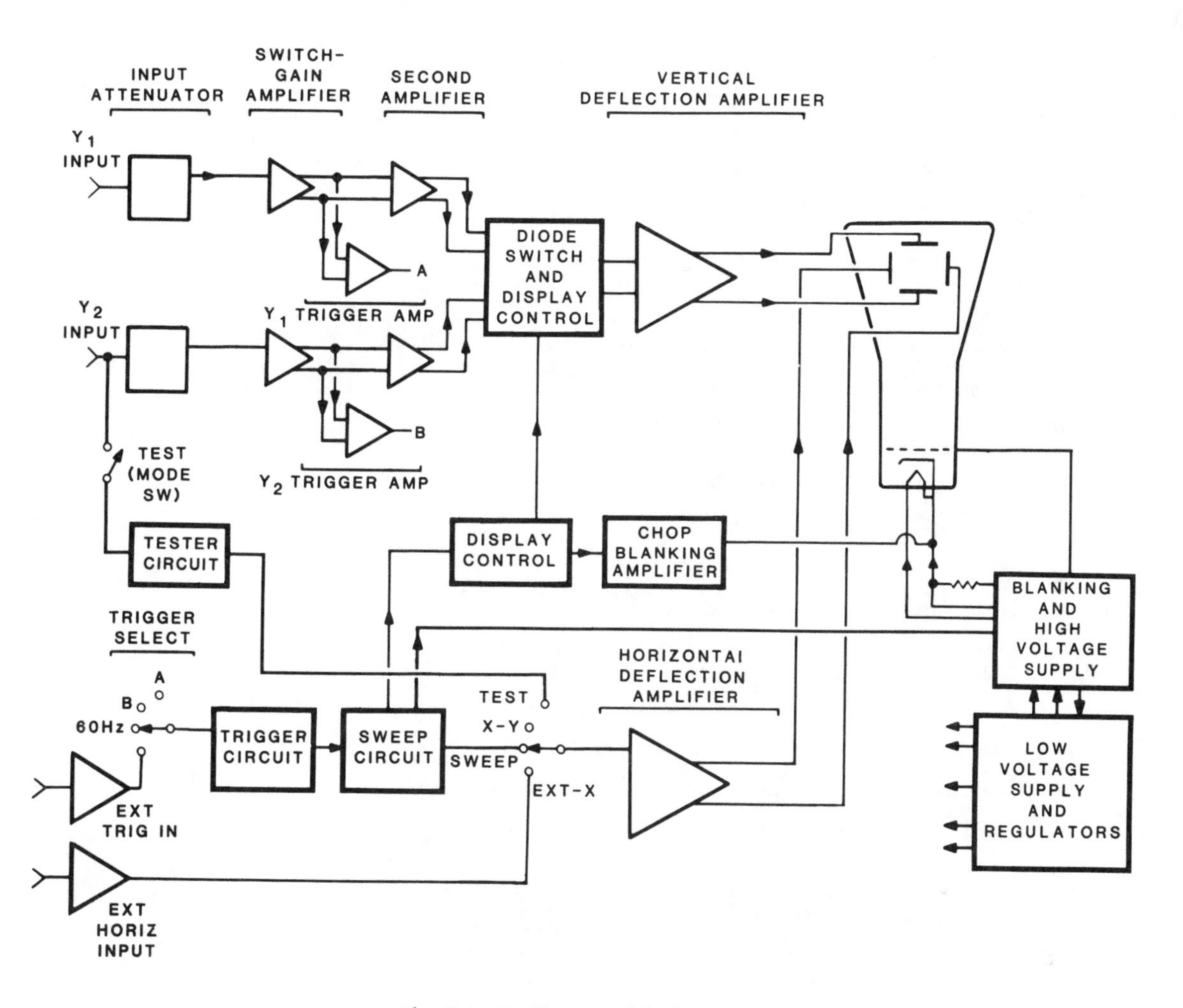

Fig. 8-1. Oscilloscope block diagram.

Vertical Circuits

The vertical preamplifier consists of two identical circuits: one for Channel Y1 and the other for Channel Y2. Components in the Channel Y1 vertical preamplifier circuit are designated by a -1 suffix, while those in the Channel Y2 vertical preamplifier are designated by a -2 suffix. For example, a Channel Y1 divider resistor is R101-1; the same resistor in the Y2 Channel is R101-2. Components without a suffix do not relate to a specific channel. Because both channels are identical, only Channel Y1 is described in the following paragraphs.

Input Circuit. When Y1 input switch SW1 (AC-GND-DC) is in the DC position, a signal applied to the Y1 input connector is coupled to the input attenuator.

Fig. 8-2. Oscilloscope schematic diagram.

SCHEMATIC OF THE
HEATHKIT®
DUAL TRACE OSCILLOSCOPE
MODEL IO-4210

NOTES:

1. ALL CAPACITOR VALUES ARE IN μF UNLESS OTHERWISE SPECIFIED.
2. ON THE HORIZONTAL CIRCUIT BOARD, RESISTORS ARE 1/4-WATT, 5% TOLERANCE UNLESS OTHERWISE MARKED. ALL OTHER RESISTORS ARE 1/2-WATT, 5% TOLERANCE UNLESS OTHERWISE MARKED. (K=1000; M=1,000,000).
3. ○ INDICATES A LETTERED WIRE CONNECTION ON A CIRCUIT BOARD.
4. ⏚ INDICATES CHASSIS GROUND.
5. ▽ INDICATES CIRCUIT BOARD GROUND.
6. ⊖ INDICATES A PART MOUNTED ON THE CHASSIS, ALTHOUGH ITS LOCATION IN THE SCHEMATIC SUGGESTS ANOTHER LOCATION.
7. CIRCUIT COMPONENT NUMBERS ARE IN THE FOLLOWING GROUPS:
   - 1 - 99 PARTS ON CHASSIS.
   - 100 - 199 PARTS ON THE VERTICAL CIRCUIT BOARD.
   - 200 - 299 PARTS ON THE HORIZONTAL CIRCUIT BOARD.
   - 300 - 399 PARTS ON THE LOW VOLTAGE CIRCUIT BOARD.
   - 400 - 499 PARTS ON THE HIGH VOLTAGE CIRCUIT BOARD.
8. ⬭ INDICATES A DC VOLTAGE MEASURED FROM THE POINT INDICATED TO GROUND WITH THE VERTICAL AMPLIFIERS BALANCED, THE TIME/CM SWITCH IN THE EXT POSITION, AND THE HORIZONTAL POSITION CONTROL CENTERED. VOLTAGE MAY VARY ± 20%.
9. TRANSISTOR TRANSISTOR LOGIC (TTL) LEVELS ARE AS FOLLOWS: A LOGIC 0, OR LOW, IS < 0.8 VOLTS. A LOGIC 1, OR HIGH, IS > 2.0 VOLTS BUT < 5.5 VOLTS.
10. ◀ THIS SYMBOL INDICATES A LINE THAT CONTINUES. TO FIND THE CONTINUED PORTION, LAY A STRAIGHT EDGE ON THE LINE THAT THE SYMBOL IS ON.
11. CRT UNBLANKED - DEPENDS ON POSITION OF INTENSITY CONTROL.
12. CRT BLANKED.
13. ONLY IN DUAL TRACE CHOPPED MODE DURING SWEEP OR EXT-X DUAL TRACE OPERATION.
14. UNIT NOT SWEEPING. SPOT CENTERED HORIZONTALLY.
15. SHADED AREAS DENOTE PARTS CRITICAL TO PRODUCT SAFETY. REPLACE ONLY WITH PARTS OF SAME TYPE AND VALUE.

✱ DEPENDS ON LINE VOLTAGE.
HO = HOLD OFF (INCLUDES RETRACE).

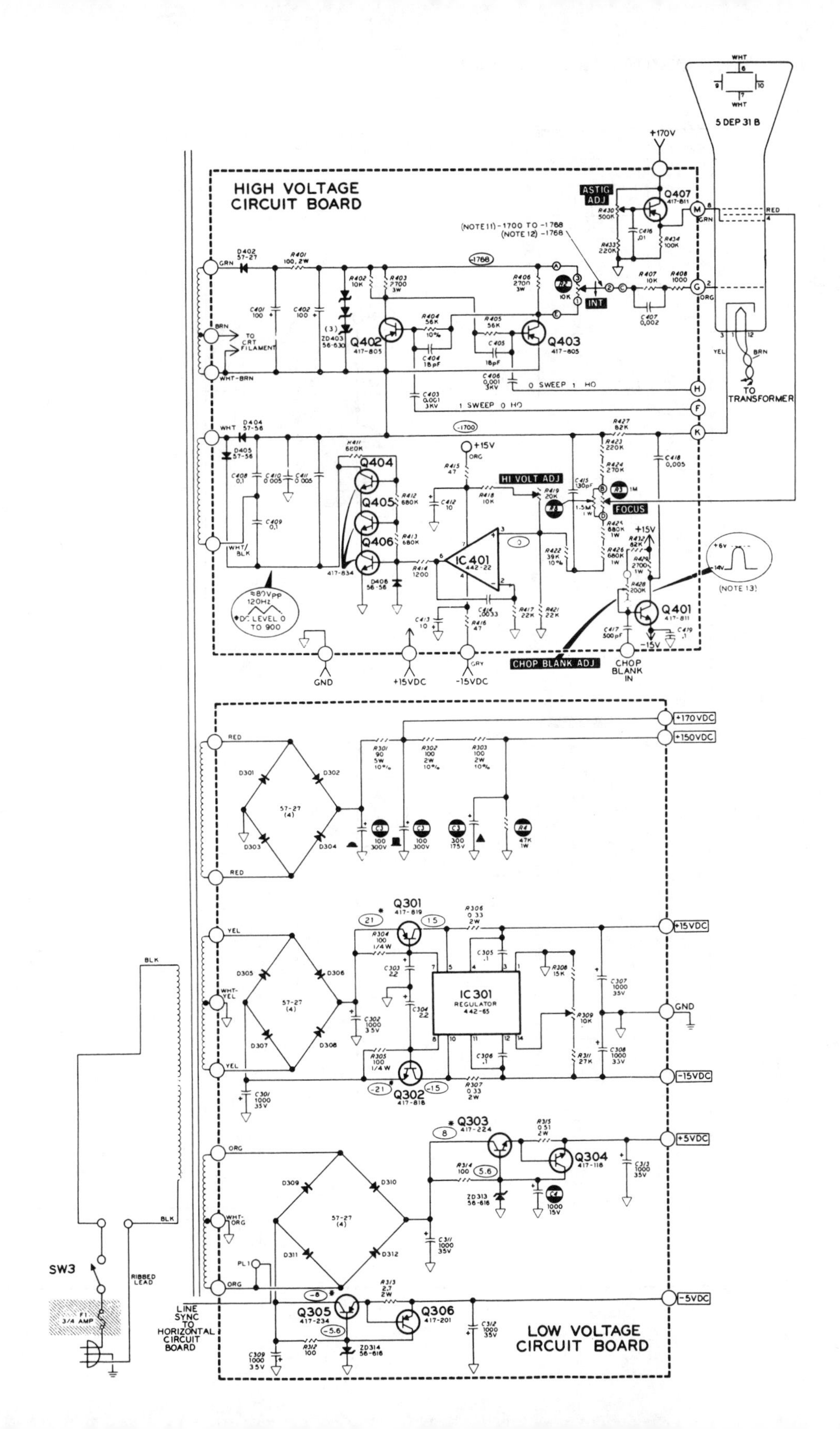

HIGH VOLTAGE CIRCUIT BOARD
LOW VOLTAGE CIRCUIT BOARD
5 DEP 31 B
+170V
ASTIG ADJ
Q407
(NOTE 11) -1700 TO -1768
(NOTE 12) -1768
INT
Q402
Q403
TO CRT FILAMENT
TO TRANSFORMER
0 SWEEP 1 HO
1 SWEEP 0 HO
Q404
Q405
Q406
IC 401
HI VOLT ADJ
FOCUS
Q401
(NOTE 13)
≈80Vpp 120Hz DC LEVEL 0 TO 900
CHOP BLANK ADJ
GND
+15VDC
-15VDC
CHOP BLANK IN
+170VDC
+150VDC
Q301
Q302
IC 301 REGULATOR 442-65
+15VDC
GND
-15VDC
Q303
Q304
+5VDC
Q305
Q306
-5VDC
SW3
F1 3/4 AMP
RIBBED LEAD
LINE SYNC TO HORIZONTAL CIRCUIT BOARD

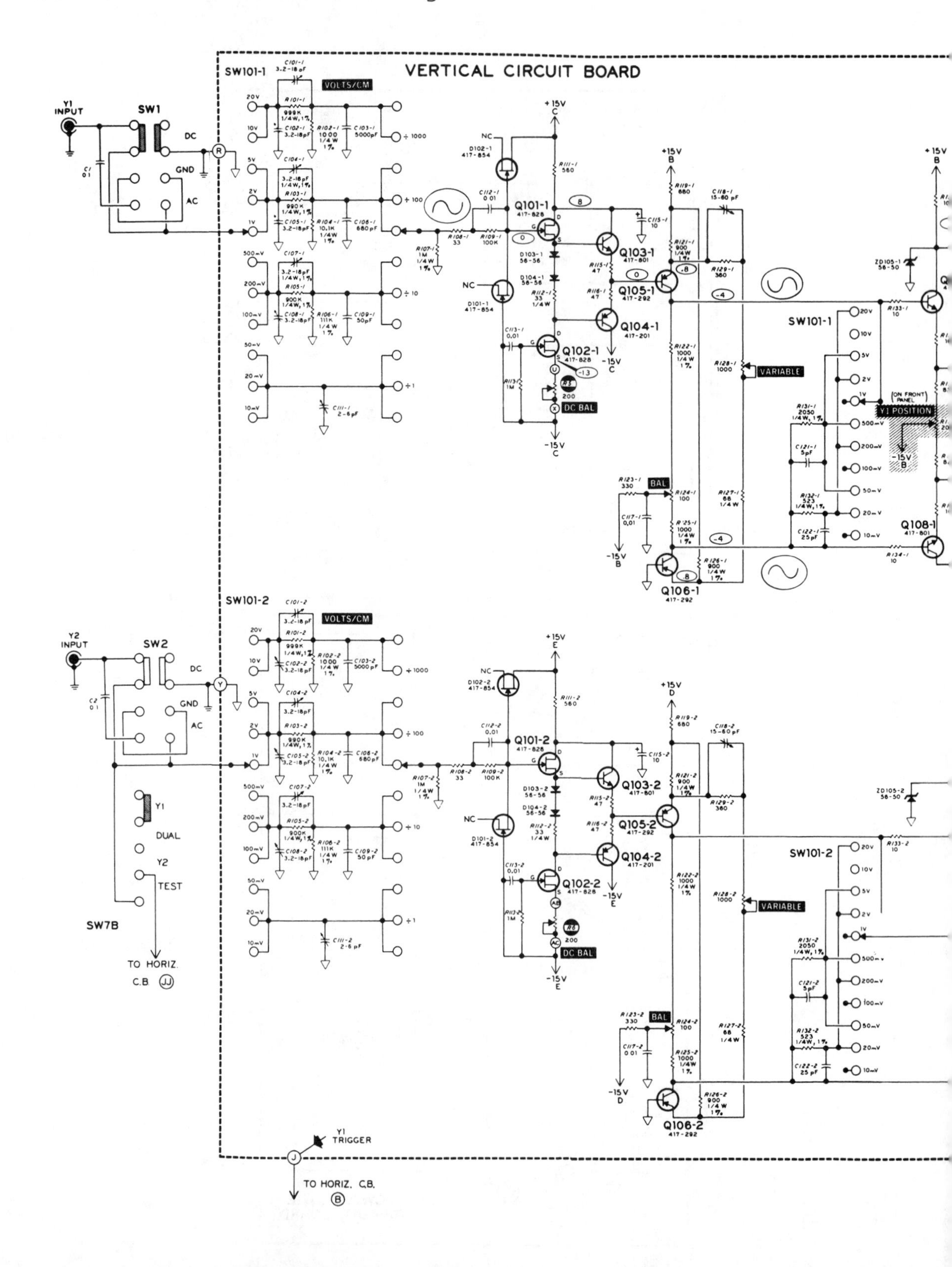
VERTICAL CIRCUIT BOARD
Y1 INPUT
SW1
DC
GND
AC
SW101-1
VOLTS/CM
Q101-1
Q102-1
Q103-1
Q104-1
Q105-1
Q106-1
Q108-1
DC BAL
BAL
VARIABLE
Y1 POSITION
(ON FRONT PANEL)
Y2 INPUT
SW2
SW101-2
Q101-2
Q102-2
Q103-2
Q104-2
Q105-2
Q106-2
Y1
DUAL
Y2
TEST
SW7B
TO HORIZ. C.B.
Y1 TRIGGER
TO HORIZ. C.B.

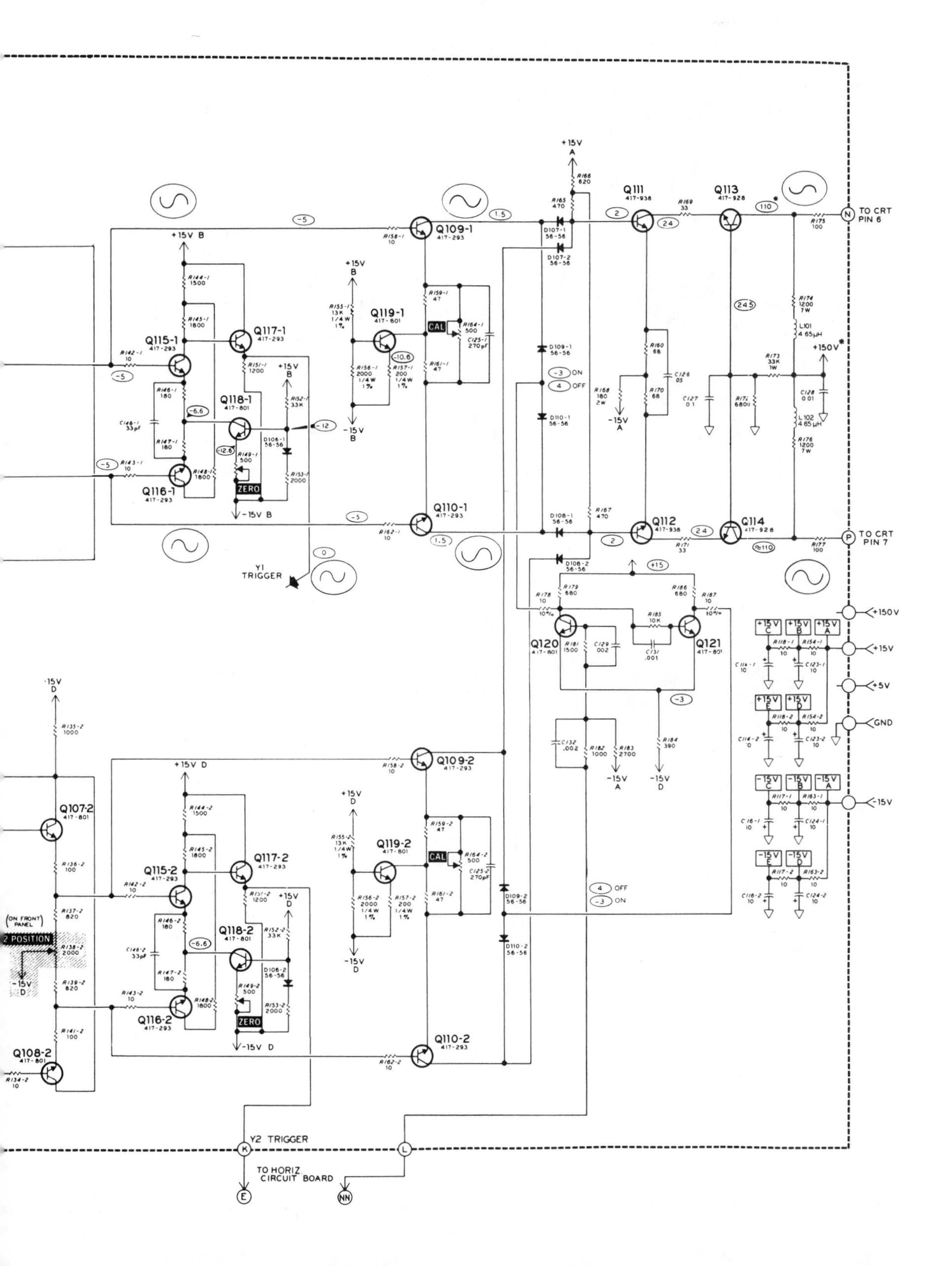

+15V A
Q111
Q113
TO CRT PIN 6
Q109-1
Q119-1
CAL
Q115-1
Q117-1
Q118-1
Q116-1
ZERO
-15V B
Q110-1
Q112
Q114
TO CRT PIN 7
+150V
Y1 TRIGGER
Q120
Q121
+15V
+5V
GND
-15V
ON
OFF
Q107-2
Q109-2
Q119-2
Q115-2
Q117-2
Q118-2
Q116-2
Q110-2
Q108-2
(ON FRONT PANEL)
Y2 POSITION
-15V D
+15V D
Y2 TRIGGER
TO HORIZ. CIRCUIT BOARD

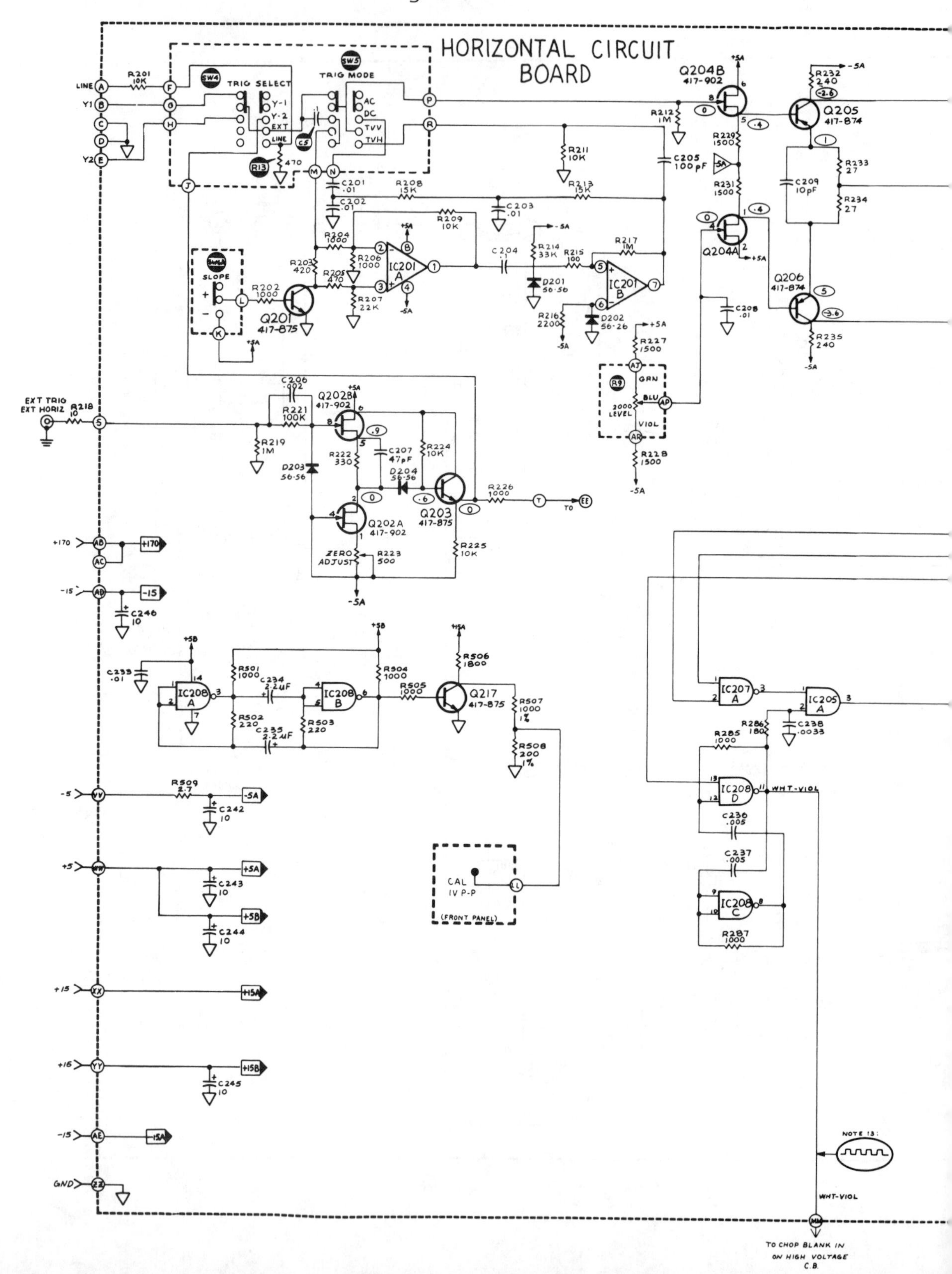
HORIZONTAL CIRCUIT BOARD
TRIG SELECT
TRIG MODE
SLOPE
EXT TRIG
EXT HORIZ
CAL
1V P-P
(FRONT PANEL)
TO CHOP BLANK IN ON HIGH VOLTAGE C.B.
WHT-VIOL
NOTE 13:

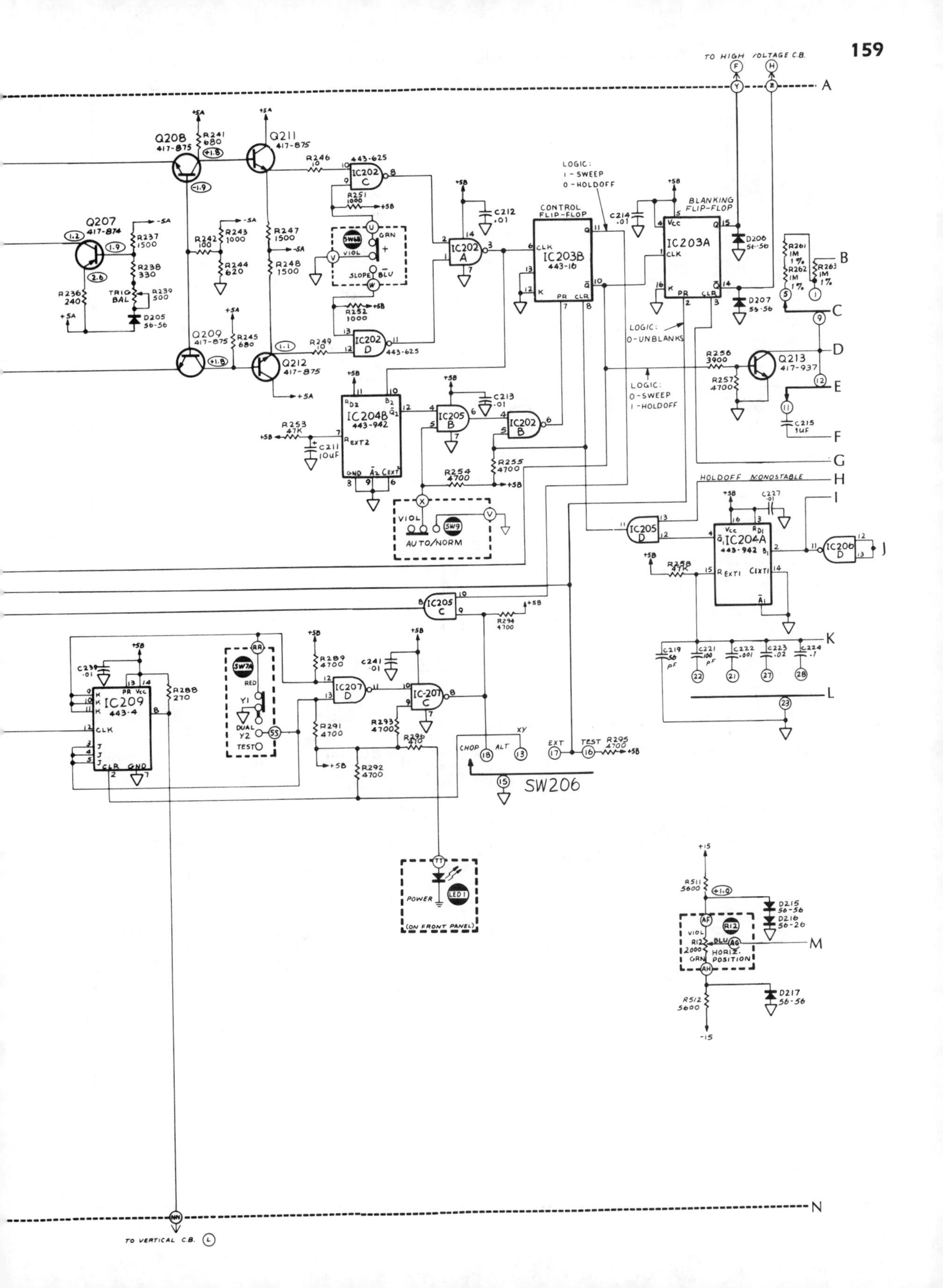
TO HIGH VOLTAGE C.B.
Q208 417-875
Q211 417-875
Q207 417-874
Q209 417-875
Q212 417-875
Q213 417-937
IC202 C
IC202 A
IC202 D
IC202 B
IC203B 443-16
IC203A
CONTROL FLIP-FLOP
BLANKING FLIP-FLOP
LOGIC: 1 - SWEEP 0 - HOLDOFF
LOGIC: 0-UNBLANKS
LOGIC: 0-SWEEP 1-HOLDOFF
IC204B 443-942
IC204A 443-942
IC205 B
IC205 C
IC205 D
IC206 D
HOLDOFF MONOSTABLE
SLOPE
TRIG BAL
AUTO/NORM
IC209 443-4
IC207 D
IC-207 C
RED
Y1
DUAL Y2
TEST
CHOP
ALT
XY
EXT
TEST
SW206
POWER
(ON FRONT PANEL)
HORIZ. POSITION
TO VERTICAL C.B.

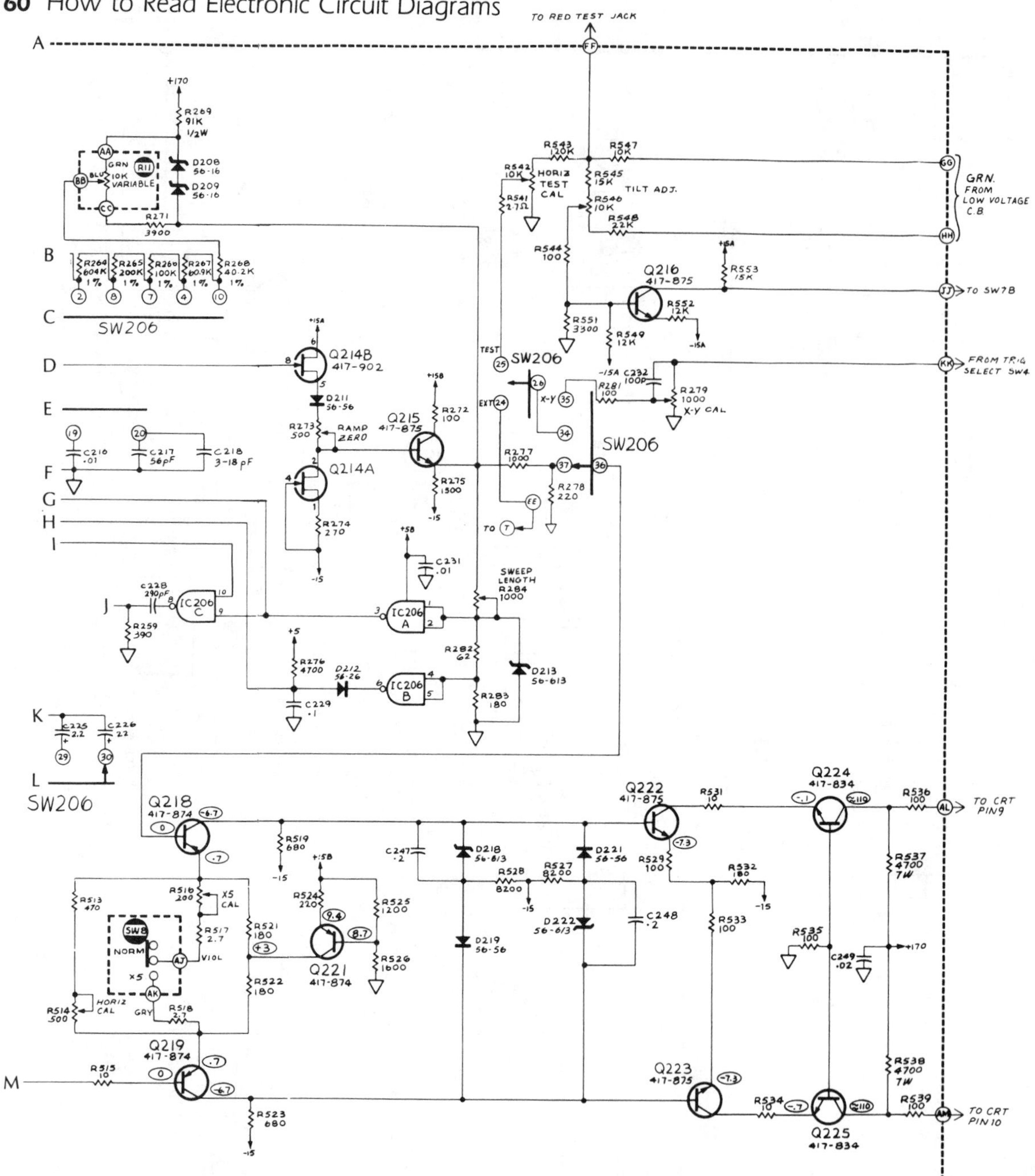

TO RED TEST JACK
A
FF
+170
R269 91K 1/2W
AA
GRN
R11
10K VARIABLE
BLU
BB
CC
D208 56-16
D209 56-16
R271 3900
B
R264 604K 1%
R265 200K 1%
R260 100K 1%
R267 60.9K 1%
R268 40.2K 1%
2
8
7
4
10
C
SW206
D
Q214B 417-902
D211 56-56
E
R273 500
RAMP ZERO
Q215 417-875
R272 100
+15A
+15B
F
19
20
C216 .01
C217 56pF
C218 3-18 pF
G
H
I
Q214A
R275 1500
-15
R274 270
-15
J
C228 290pF
IC206 C
R259 390
IC206 A
+58
C231 .01
SWEEP LENGTH R284 1000
+5
R276 4700
D212 56-26
IC206 B
C229 .1
R282 62
R283 180
D213 56-613
K
C225 2.2
C226 22
29
30
L
SW206
R543 120K
R547 10K
R542 10K
HORIZ TEST CAL
R541 2.7Ω
R545 15K
TILT ADJ.
R546 10K
R548 22K
GG
HH
GRN. FROM LOW VOLTAGE C.B.
R544 100
Q216 417-875
R553 15K
JJ
TO SW7B
R551 3300
R552 12K
R549 12K
-15A
TEST
25
SW206
26
-15A
C232 100P
KK
FROM TRIG SELECT SW4
R281 100
R279 1000
X-Y CAL
EXT 24
X-Y 35
34
SW206
R277 1000
37
36
R278 220
EE
TO 7
Q218 417-874
Q222 417-875
R531 10
Q224 417-834
R536 100
AL
TO CRT PIN9
R519 680
C247 .2
D218 56-613
D221 56-56
R529 100
R528 8200
R527 8200
R532 180
R537 4700 7W
R516 200
X5 CAL
R524 220
R525 1200
R513 470
SW8
NORM
AJ
R517 2.7
VIOL
R521 180
Q221 417-874
R526 1600
D219 56-56
D222 56-613
C248 .2
R533 100
R535 100
C249 .02
+170
X5
AK
R522 180
R514 500
HORIZ CAL
GRY
R518 2.7
Q219 417-874
M
R515 10
R523 680
Q223 417-875
R534 10
Q225 417-834
R538 4700 7W
R539 100
AM
TO CRT PIN 10
N

When the Y1 input switch, SW1, is in the AC position, this signal is coupled through capacitor C1, which passes only ac signals. This permits an ac signal, superimposed on a dc potential, to be seen without the dc component being displayed. The GND position of this switch disconnects the input signal and grounds the attenuator input. This allows the trace to be adjusted to a zero reference without disconnecting the test leads from the circuit under test.

Because the second and vertical deflection amplifiers (Q109-1/Q110-1 and Q111-Q114, respectively, which are discussed later) operate at a fixed gain, any signal applied to them must be within a useable range, approximately 80 mV/cm. Therefore, the primary purpose of the vertical input circuits is to reduce or increase the input signal by a known factor to this useable level.

The vertical input circuit basically consists of an attenuator, an input follower, and a switched-gain amplifier. These circuits function together, through the VOLTS/CM switch, to provide the total desired attenuation or gain. The attenuator obtains its four attenuation factors (1, 10, 100, and 1000) from four divider networks, resistors R101-1 through R106-1, and capacitors (C101-1, C103-1, C104-1, C106-1, C107-1, and C109-1.) At dc and low ac frequencies, the resistive dividers reduce the input signal level, while at higher frequencies, attenuation is determined by the resistor-capacitor (RC) networks.

Trimmer capacitors C101-1, C104-1, and C107-1 are used to adjust the capacitor division ratio to match the resistor ratio. Trimmer capacitors C102-1, C105-1, and C108-1, and C111-1 are adjusted during calibration to make the input capacitance of the oscilloscope equal on all positions of the VOLTS/CM switch. This is essential when an attenuation probe, usually X10, is used.

The input-follower circuit consists of a FET source follower, dc current source, and an impedance translator. The attenuated input signal is coupled through resistors R108-1 and R109-1 and capacitor C112-1 to the gate of FET source follower, Q101-1. Capacitor C112-1 forms a high-frequency path around R109-1 for improved frequency response. Input protection is provided by two FETs, D101-1 and D101-2, wired as reverse-biased diodes. They are connected to the positive (+) and negative (−) 15-V supplies. Thus, if the input signal following the input attenuator exceeds 15 V, the FETs become forward biased and limit the signal to within a diode drop of 15 V. This prevents damage to Q101-1 if the VOLTS/CM switch is in a low range and a high potential is applied to the input.

Transistor Q101-1 provides the high-input impedance necessary to prevent attenuator loading and a low output impedance to drive emitter-follower transistors Q103-1 and Q104-1. To compensate for the dc voltage present at the source of Q101-1 when no signal is applied, FET Q102-1 forms a dc current source. Dc BAL control R5 is adjusted so that the current supplied is sufficient to produce a zero output at the source (S) of Q101-1 for a zero input at its gate (G). The circuit formed by diodes D103-1 and Q104-1 and transistors Q103-1 and Q104-1 acts as an impedance translator. It reduces the output impedance of the input follower to approximately 50Ω. The output of the input follower is coupled to the switched-gain amplifier.

The switched-gain amplifier is formed by transistors Q105-1 and Q106-1 to provide a double-ended output from a single-ended input signal. A relatively constant current is supplied through resistor R119-1 to the amplifier, so an increase in current through Q105-1 will cause a corresponding decease in current through Q106-1. Thus, as Q105-1 amplifies the input signal, Q106-1 produces an equal, but opposite, signal. This creates a push-pull effect on the signal, which is amplified in the following stages to drive the CRT vertical deflection plates. Front panel VARIABLE control R128-1 adjusts the gain of the amplifier when it is turned from its detented CAL (fully clockwise) position.

Two switch-selected RC networks reduce the gain of this switched-gain amplifier from 8 to 4 and 1.6. Table 8-1 shows how the VOLTS/CM switch selects the various attenuation factors and gains of the switched-gain amplifier to provide the desired total gain. Step BALANCE control R124-1 adjusts the collector currents of Q105-1 and Q106-1 so that the CRT trace does not shift when the gain through the VOLTS/CM switch is changed.

Volts/CM Position	Attenuation Factor	Amplifier Gain	Total Gain Factor
20 V	÷1000	4	.004
10 V	÷1000	8	.008
5 V	÷100	1.6	.016
2 V	÷100	4	.04
1 V	÷100	8	.08
500 mV	÷10	1.6	.16
200 mV	÷10	4	.4
100 mV	÷10	8	.8
50 mV	÷1	1.6	1.6
20 mV	÷1	4	4.0
10 mV	÷1	8	8.0

Table 8-1. Gain Factors for the IO-4210.

To illustrate how the attenuator and switched-gain amplifier work together for the proper gain, assume the VOLTS/CM switch is in the 10 mV position and a 10 mV signal is applied to the input. Because the total gain factor is 8, the input signal is amplified by a factor of eight before it is coupled to follower Q107-1/Q108-1. An 80 mV signal at the follower will cause a 1 cm deflection on the CRT. Now assume the VOLTS/CM switch is in the 500 mV position and the 500 mV signal is applied to the input. The total gain factor is now .16. Multiplying the input signal level by the total gain factor results in an 80 mV signal to the follower (500 mV × .16 = 80 mV), again causing a 1 cm deflection on the CRT.

Differential emitter-follower Q107-1/Q108-1 serves as a buffer between the switched-gain amplifier and the second amplifier. It also provides vertical trace positioning channel. Y1 POSITION control R138-1 controls the trace position by shifting the emitter current between the two emitter circuits. The signal is coupled from the follower to the trigger amplifier and second amplifier. The trigger amplifier is described, following the Vertical Deflection section.

The second amplifier is a differential amplifier with a gain of approximately 10. Its output is direct-coupled to the diode bridge. CAL control R164-1 adjusts the gain of this amplifier and the overall calibration of the vertical circuit.

Diode Switch. Both preamplifier circuits, Channels Y1 and Y2, share the vertical deflection amplifier. This is accomplished with two high-speed diode switch networks, D107-1 through D110-1 and D107-2 through D110-2, that are actuated by the display control circuit. When one diode switch is turned on, the other is turned off so that only one signal can be coupled to the vertical deflection amplifier. Two-channel operation is accomplished by turning each diode-switch network on and off at a rapid rate or on alternate display sweeps. Control of the diode switch is described later.

Vertical Deflection. From the diode switch, the input signal is direct-coupled to the vertical deflection amplifier. This amplifier, consisting of transistors Q111-Q114, is wired in a differential cascade configuration, with a gain of approximately 20. Capacitor C126, across the emitters of Q111 and Q112, provides high-frequency square-wave compensation. Ferrite beads, FB101 and FB102, in the common-base amplifiers Q113/Q114, prevent oscillations in the amplifier. Circuit loading is supplied by resistors R174 and R176 and inductors L101 and L102 serve as peaking coils. The output of this amplifier is coupled to the vertical deflection plates of the CRT for beam control. Vertical beam deflection requires between 12 V and 15 V/cm, depending on individual CRT requirements. Vertical CAL control R164-1 in the vertical input circuit adjusts overall vertical gain to match the CRT deflection characteristics.

Trigger Amplifier. A differential amplifier and follower comprise the trigger amplifier circuit. Its output is used to supply a trigger signal to the horizontal time base, trigger, and sweep circuits. In addition, the Channel Y1 trigger signal can be switched to the horizontal deflection circuit for X-Y operation.

A portion of the input signal is coupled from follower Q107-1/Q108-1 to the input of the differential amplifier in the trigger amplifier circuit. Emitter follower Q117-1 couples the trigger signal from the inverting leg (Q115-1 of the differential amplifier) to the horizontal time base, trigger, and sweep circuits. Transistor Q118-1 is a temperature-compensated constant current source for this circuit. Zero control R149-1 (in the emitter leg) adjusts the current so the output of the follower will be zero with no signal input to the trigger amplifier. Thus, the circuit performs as a differential to the single-ended converter.

Trigger, Sweep, and Control Circuits

On command from a trigger pulse, the horizontal time-base circuits generate a linear-ramp sweep signal to drive the CRT horizontal deflection plates and move the dot across the screen at a constant rate. In the automatic-triggering mode of operation, if no trigger is present, the time-base circuits free-run and generate an auto-baseline. An oscillator provides the vertical signal chop rate when two traces are displayed in the chop mode, in which the TIME/CM sweep rate is 5 ms and slower. Figure 8-3 shows a typical two-trace, chopped display that has been exaggerated for clarity.

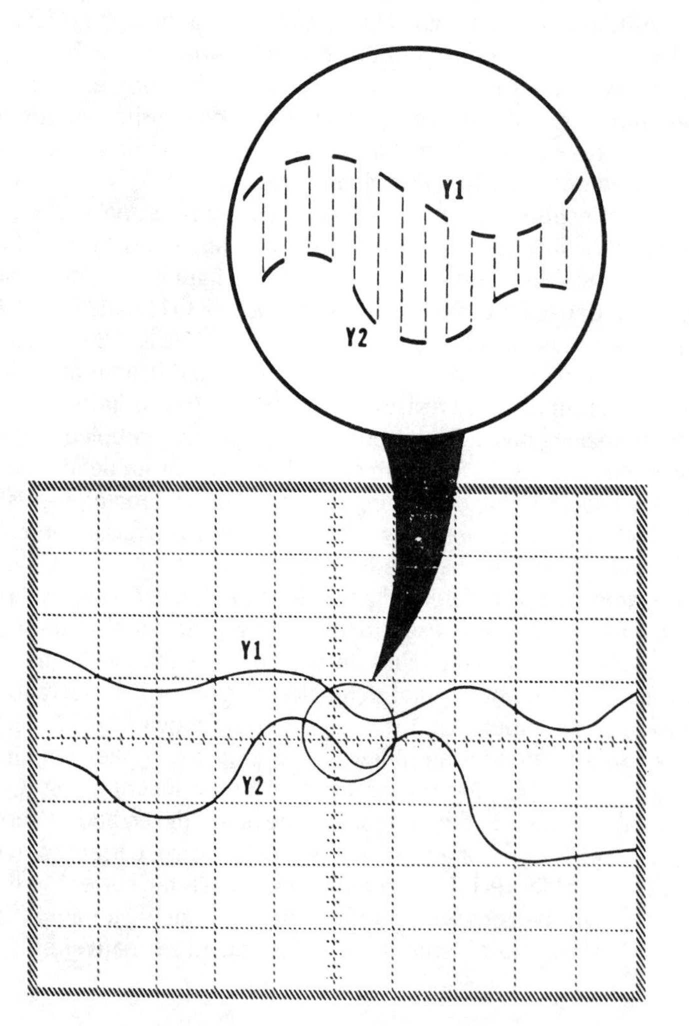

Fig. 8-3. Typical two-trace chopped display.

When a trigger pulse of sufficient amplitude is present, the trigger signal passes through the slope selector gate. The signal from the gate turns on the sweep control and allows the time capacitor to charge through the "bootstrap" constant-current source. The charging of the capacitor produces a linear ramp signal that is coupled through the voltage follower to the horizontal deflection circuits.

The ramp signal is also coupled to the sweep-end circuit. When the ramp reaches a preset voltage level, set by the Sweep Length control, the sweep-end circuit triggers the blanking flip-flop and sweep-end monostable multivibrator. The sweep-end monostable circuit ensures that retrace will not occur until after the CRT has been blanked.

Trigger. In the automatic triggering mode, the trigger circuit examines the trigger signal for a proper trigger point. If the signal is large enough, the sweep circuit is activated by the trigger. If the signal is insufficient or absent, the sweep circuits are allowed to free-run.

Depending on the desired trigger mode, one of four sources can be selected by the TRIG switch, EXTERNAL TRIGGER, CHANNEL Y1 TRIGGER, CHANNEL Y2 TRIGGER, or LINE SYNC. The Channels Y1 and Y2 TRIGGER signals are provided by the vertical preamplifier trigger circuits, while the LINE SYNC signal is tapped directly off one side of the power transformer's 6-V winding.

The external trigger amplifier consists of source follower Q202B, constant current source Q202A, and emitter-follower Q203. Resistor R219 sets the input impedance, while resistor R221 and diode D203 provide overvoltage protection. Capacitor C206 provides a high-frequency path around R221. Resistor R223 sets the bias voltage so the emitter of Q203 can be set at zero volts. Resistor R226 provides decoupling and applies the output of the amplifier to the horizontal amplifier when an external horizontal signal is selected.

The trigger signal is coupled from TRIG SELECT switch SW4 to TRIG MODE switch SW5. When you select AC, DC, TV-H, transistor Q201, integrated circuits IC201A, IC201B, and their associated circuits form a sync separator to provide stable triggering on composite video signals. When TV-V or TV-H is the selected input, trigger signals are processed by the sync separator before they are routed to the trigger amplifier. In normal operation, the input to the sync separator is a composite video signal. The sync separator removes the video information and separates the vertical and horizontal sync pulses.

Transistor Q201 and switch SW6A allow selection of triggering from a signal's positive or negative slopes. Integrated circuit U201 serves as an amplifier for either of these signals and has an approximate gain of 10. When the slope is selected, Q201 is turned on by SW6A. Resistors R204 and R206 form a 2:1 divider for the input signal. When the (+) slope is selected, the full signal is applied to the noninverting input pin 3 of IC201A. The resulting signal is approximately twice the signal supplied by the input at pin 2 of the IC and cancels the inverted signal at that point.

Capacitor C204 couples the composite video signal to diode D201, which clamps the negative-going sync pulses to −.7 V. The output is then routed to the positive input of IC201B. The negative input of this IC201B is held at −.3 V by diode D202. The output of comparator IC201B is high whenever the composite video signal on pin 5 is more positive than −.3 volts; only the horizontal and vertical pulses can make pin 5 more positive than −.3 volts. Therefore, whenever a horizontal or vertical sync pulse occurs, the output of IC201B goes low. This makes the output of IC201B a series of negative-going pulses synchronized by the horizontal or vertical sync pulses. Capacitor C205 and resistor R211 form a high-pass filter that passes only the horizontal sync pulses to switch SW5. The low-pass filter, formed by resistors R208 and R213, along with capacitors C201 through C203, couple the vertical sync pulses to switch SW5.

The trigger signal is applied to the gate (pin 8) of transistor Q204B. Transistor Q204A is the current source for Q204B, while the gate bias is controlled by resistor R9 which sets the desired trigger input level. Transistors Q205, Q206, Q208, and Q209, comprise a cascode amplifier in which transistor Q207 provides a constant current source. Resistor R239 can be adjusted to balance this differential stage. Transistors Q211 and Q212 are followers that drive the slope detector gate, composed of ICs 202C and 202D. SLOPE switch SW6B selects the positive or negative trigger slope. When this switch is in the positive (+) position, a high is applied to pin 13 of IC202D allowing any negative-going pulse applied to pin 12 to be coupled through. At the same time, a low is applied to pin 10 of IC202C which locks the output of the device high. When the negative (−) slope is selected by SW6B, a low is applied to pin 13 of IC202D, locking its output high. At the same time, a high is applied to pin 9 of IC202C, allowing pin 10 to control its output. IC202A serves as a buffer and an inverter to drive control flip-flop IC202B, and to turn off the auto-baseline generator IC204B when there is a trigger signal of sufficient amplitude.

Sweep. The negative edge of the trigger pulse activates IC203B and turns transistor Q213 off. This lets the timing capacitor charge through the bootstrap current source and generates a linear voltage ramp. The ramp, or sweep signal, is coupled to the horizontal deflection circuit and the remaining sweep circuits. When the ramp reaches a predetermined level, the CRT is blanked, IC203B is reset, Q213 is turned on to provide a discharge path for the timing capacitor, and the ramp returns to zero. Refer to Fig. 8-4 as you read the following information.

When transistor Q213 turns off, the timing capacitor begins to charge. After approximately 10 ns, the CRT unblanks and the trace becomes visible. This short delay hides any switching transients. At a preset ramp level, the CRT is again blanked before the trace is stopped, to give the CRT time to fully blank. After the short delay, transistor Q213 is turned on and the timing capacitor is discharged.

Initially, sweep control IC203B is in a reset condition, Q is low and $\overline{Q}$ is high, and transistor Q213 is turned on. A trigger pulse from slope-selector gate IC202 will toggle IC203B and switch Q high and $\overline{Q}$ low. The low $\overline{Q}$ turns transistor Q213 off and toggles blanking control IC203A. As Q213 turns off, the timing capacitor begins to charge through the bootstrap current source. At the same time, IC203A switches the CRT blanking circuit and unblanks the CRT.

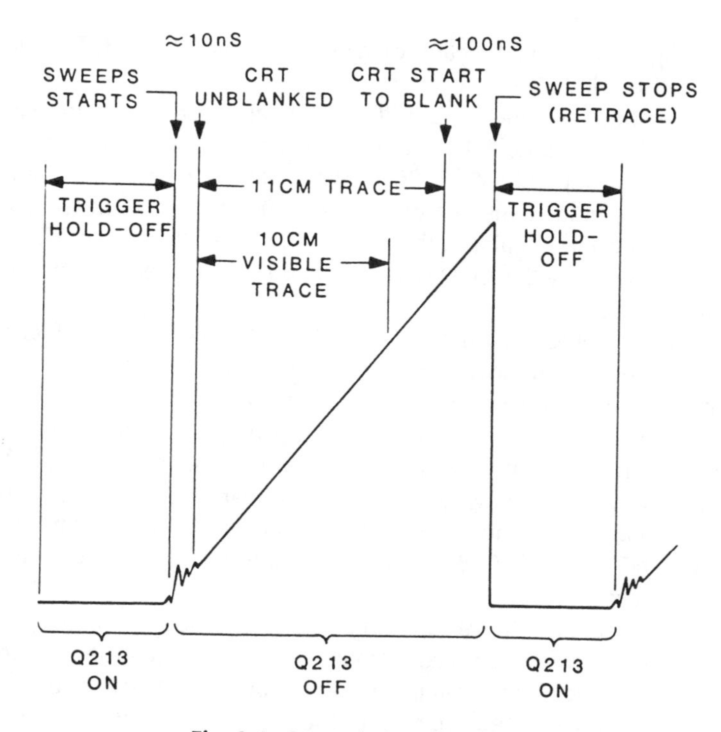

Fig. 8-4. Q213 timing diagram.

The bootstrap current source is part of the sweep generator. FETs Q214A and Q214B, with transistor Q215, form a voltage follower with a gain of approximately 1. It has a very high input impedance to prevent circuit loading, which could cause a nonlinear voltage sweep ramp. The junction of resistors R269 and R11 are held to a level 10 V above the output of the follower by 5-V zener diodes D208 and D209. Because the follower input voltage equals the output voltage, the voltage across the selected timing resistor will always be constant. This will produce a constant current to charge the selected timing capacitor. When VARIABLE control R11 is turned from its CAL position, the voltage differential is lowered. Thus the charging current will be reduced and, as a result, reduce the sweep speed. The variable control is used to provide continuous sweep speeds between calibrated ranges. RAMP ZERO control R273 adjusts the follower for proper voltage offset.

The output of the sweep generator is coupled through voltage divider R277 and R278 to the horizontal deflection amplifier. It is also coupled through SWEEP LENGTH control R284 to sweep and Schmitt gate IC206A. SWEEP LENGTH control R284 is adjusted so the output of the Schmitt trigger gate will go low when the ramp voltage exceeds approximately 1.6 V. This represents a horizontal sweep of approximately 11 cm. The low from IC206A resets, or clears, blanking control IC203A, which blanks the CRT. Zener diode D213 protects the Schmitt trigger from misalignment or malfunction of sweep generator IC206B. The function of IC206B will be described later.

The low from sweep-end gate IC206A is also coupled to the sweep-end monostable IC, 206C and D. This converts the low level signal to a short-duration negative pulse of approximately 100 ns, which gives the CRT time to fully blank, and couples it to hold-off monostable IC204A. The hold-off monostable IC toggles and Q goes low and remains in this condition until it "times out." The hold-off time is determined by the TIME/CM switch, and is of sufficient duration to ensure complete retrace.

With pin 13 of IC205D high, from IC206B, the low from $\overline{Q}$ of IC204A will force pin 11 of IC205D low and reset sweep control IC203B. This forces the $\overline{Q}$ output high and turns on Q213, which quickly discharges the sweep timing capacitor. The low from IC205D pin 11 does not affect gate IC202B because IC202A pin 4 is already low due to the auto-baseline monostable IC204B, toggled by the trigger signal. The low from hold-off monostable IC204A "locks up" sweep control IC203B so it cannot toggle on a trigger signal until after hold-off. After IC204A "times out," IC203B can toggle on the next trigger signal and start a new sweep cycle.

If, for any reason, the sweep control circuitry should "hang up," such as at initial turn-on, the ramp voltage would continue to increase. A high voltage level would activate IC206B, forcing the output of IC205D high, resetting IC203B, and discharging the sweep timing capacitor to initiate a new cycle.

Normally, recurring trigger pulses hold monostable IC204B on. The low at the $\overline{Q}$ output is coupled through IC205B and holds the output of IC202B high for normal sweeps to take place. However, with no input-trigger pulses, IC204B times out and its $\overline{Q}$ output goes high. This, through IC204B, forces the output of the IC202B low, allowing IC203B to free-run and produce sweeps for a baseline. In the NORMAL mode of operation, the automatic baseline will never appear, as this feature is overridden by switch SW9.

Control. For single-channel operation, the diode switch on the vertical circuit board is latched by the diode switch control circuit to couple the selected input to the vertical deflection circuit. When an input channel is selected by SW7A, a logic high is placed on the J or K input, depending on which channel is turned on, of display control IC209. The first low-to-high level transition at the $\overline{Q}$ output of IC203B is coupled to pin W of IC207A. With a high on pin 1 of IC207A, the signal on pin 3 is inverted and coupled through IC205A. Here pin 2 goes high, clocking input pin 12 of IC209. The high-to-low level transition toggles IC209 and couples the information at the J input to the Q output. Thereafter, each clock pulse will have no effect while the J-input level does not change. The Q output, high for Y1 and low for Y2, is coupled to the input of the diode switch control.

A logic high to the diode switch control will turn transistor Q120 on and reverse bias diodes D109-1 and D110-1. Diodes D107-1 and D108-1 are then forward biased and the signal for Channel Y1 is coupled to the vertical deflection amplifier. With transistor Q120 turned on, Q121 is turned off. Diodes D109-2 and D110-2 are forward biased and D107-2 and D108-2 are reverse biased, which blocks the Channel Y2 signal.

Dual-channel operation is permitted when both input channels are enabled by switch SW7A. The TIME/CM switch selects either the chop mode, from .2 sec to 2 mS, or the alternate channel mode, from 2 mS to .2 μS. When both Y1 and Y2 channels are turned on, logic highs are coupled to the J and K inputs of display control IC209. IC207D is also enabled to put IC207C pin 8 high.

In the ALTERNATE channel mode, IC205C pin 9 is grounded (low) by the TIME/CM switch, which disables IC205C and chopper oscillator IC208C/IC208D. At the beginning of a sweep, the high-to-low level transition at $\overline{Q}$ of IC203B is inverted by IC207A whose pin 1 is high and coupled through pin 2 of IC205A which is high to U209 pin 12. This sets the display control. At the end of the sweep, the low-to-high level transistor from the $\overline{Q}$ output of IC203B toggles IC209 and the Q output changes state, either from low to high or from high to low. Output Q is coupled to the diode switch control circuit. A logic high will couple the Channel Y1 signal to the vertical input circuit, while a logic low will couple Channel Y2. Each successive sweep cycle will toggle IC209 and alternate the channel being coupled through the diode switch to the vertical deflection circuits.

For CHOP mode operation, an oscillator provides a switching signal to the display control. This forces the diode switch to alternately couple Channels Y1 and Y2 to the vertical deflection circuit many times during each sweep. Because the switching rate of approximately 200 kHz is so fast and sweep rate is comparatively slow, the chopped effect of the display is not visible. IC205C pin 9 is high, which enables IC205C. When a new sweep cycle begins, the Q output of IC203B goes high. This high is coupled through IC205C to chopper oscillator IC208C/IC208D pin 13 and turns the oscillator on. The low from $\overline{Q}$ of IC203B is inverted by IC207A to enable IC205A. Therefore, each low-high-low pulse from the chopper oscillator will be coupled through resistor R286 and IC205A to the clock input of IC209. Each clock pulse is also coupled to the chop-blanking amplifier. The short time delay caused by the R286/C238 RC network ensures that the CRT will be blanked when the diode switches from one input channel to the other. This serves two functions. First, blanking the CRT will remove any trace lines that might appear when the vertical circuits switch between input channels. Second, the short delay between blanking and switching will hide any possible switching noise that might be generated. At the end of the sweep cycle, the chopper oscillator is turned off by the Q output of IC203B. This removes the possibility of a false trigger after hold-off. A new sweep cycle will start the chopper oscillator anew.

Each chop pulse is coupled to the chop-blanking amplifier through capacitor C417 on the high-voltage circuit board. This capacitor acts as a differentiator to convert the incoming square wave to dual-polarity spikes. See Fig. 8-5. Control R428 controls the gain of transistor Q401 and holds it on until a negative spike turns Q401 off for a very short time, as determined by the value of C417. The positive portion of the spike has no effect on the circuit. When Q401 turns off, the collector voltage rises sharply to +15 V. This short pulse is coupled through capacitor C418 to effectively pull the CRT cathode voltage up to −1670 V, which will momentarily blank the CRT. Normal CRT blanking or retrace will be described with the high-voltage power supply.

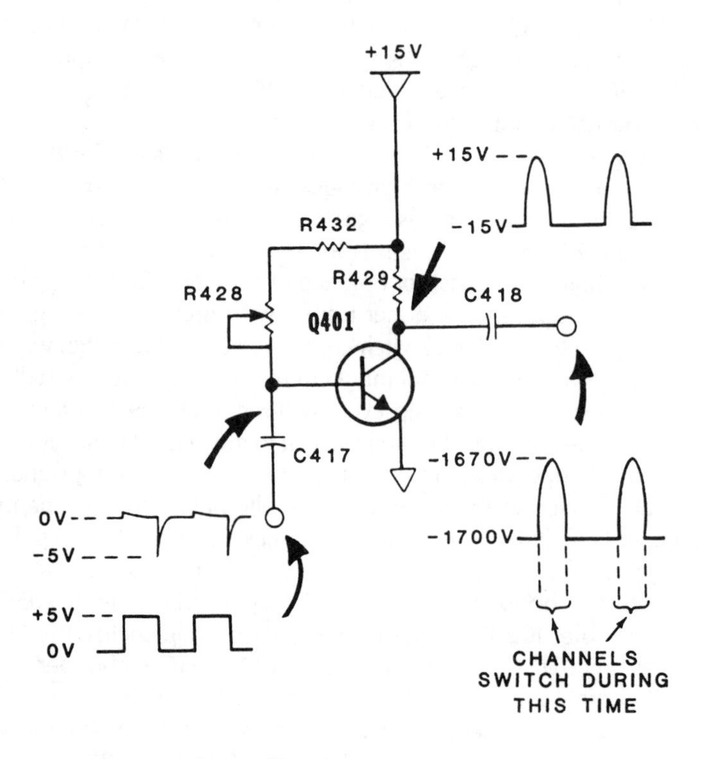

Fig. 8-5. Chop-blanking amplifier circuit.

Horizontal Circuits

The horizontal signal is coupled from the sweep generator to the base of transistor Q218 in the first horizontal deflection differential amplifier, Q218/Q219. A reference voltage from the horizontal position control provides an offset adjustment at the base of Q219. Amplifier gain is approximately 10. Control R514 adjusts the gain of this amplifier and, thereby, the overall gain of the horizontal deflection circuit. This is to compensate for the individual deflection characteristics of the CRT. Transistor Q221 is the constant current source for this amplifier section.

Final amplification occurs in the second deflection amplifier. It is a cascode differential amplifier with a gain of approximately 25. The output is coupled to the horizontal deflection plates of the CRT for beam control. Horizontal beam deflection requires between 19 and 25 V/cm, depending on individual CRT characteristics. Control R516 is used to adjust the emitter resistance of Q218/Q219 so that when switch SW8 is in the X5 position, the gain of the amplifier is multiplied by exactly 5. Resistors R526 and R527, capacitors C247 and C248, with diodes D219 through D222, form a limiter network to prevent transistors Q222 and Q223 from overdriving while they are in the X5 mode.

When X-Y operation is desired, the trigger signal from the Channel Y1 trigger amplifier is coupled to the base of Q218 through the TIME/CM switch. The TIME/CM switch also grounds the clear input to display control IC209. This forces the Q output low to ensure that the diode switch couples the Channel Y2 signal to the vertical deflection amplifier. The low from the TIME/CM switch also presets the blanking control to ensure that the CRT remains continuously unblanked. X-Y CAL control R279 is used to calibrate overall horizontal deflection gain for the X-Y mode, while capacitor C232 reduces high-frequency phase-shift.

External X operation allows you to look at two signals in relation to a third signal; in essence, you have a two-channel X-Y. With the TIME/CM switch in the EXT X position, a low is coupled to IC U203A to preset the blanking control and unblank the CRT. Because the sweep circuit has been disabled, the chopper oscillator is unlatched and runs continuously. The external sweep signal is coupled to the horizontal deflection amplifier through the external horizontal input amplifier.

Power Supplies

+170- and +150-Volt Supplies. Full-wave rectifier diodes D301 through D304 produce the +170-V supply used in the horizontal deflection amplifier and the astigmatism control circuit. A second filter divides the 170-V source down to +150 V for the vertical deflection amplifier.

±15-Volt Supply. Diodes D305 through D308 comprise two full-wave rectifiers that produce positive and negative 21 Vdc from the power transformer. These are filtered and then coupled through pass-transistors Q301 and Q302. The transistor base (B) voltages on these devices is set by dual-polarity tracking regulator IC301. Resistors R306 and R307 set the maximum current that the regulator will supply. In turn, this maximum will limit the supply current for short-circuit protection. VOLTAGE ADJ control R309 calibrates the −15-V line and, through the regulator, the +15-V line. Capacitors C307 and C308 reduce load transients.

±5-Volt Supply. Full-wave rectifier diodes D309 through D312 produce a positive and negative 8 Vdc from the power transformer. Capacitors C309 and C311 filter the 8 V, while transistors Q303 and Q305 serve as pass elements. Their outputs are limited to 5 V by zener diodes ZD313 and ZD314, which are connected to the bases of transistors. In the positive 5-V supply, resistor R315 sets the base-to-emitter (B-E) current of Q304 to shut down the supply when it exceeds approximately 400 mA. Resistor R313 and transistor Q306 limit the negative 5-V supply to approximately 100 mA. Capacitors C312 and C313 reduce load transients.

The line-sync trigger signal is coupled from the 8-V transformer secondary winding through resistor R206 to the TRIG switch. The pilot lamp and resistor R7 are also connected across the 8-V secondary winding.

High Voltage. Diodes D404 and D405, with capacitors C408 through C411, comprise a voltage doubler that produces approximately −2000 Vdc at a nominal line voltage. The positive output of the doubler is connected to the collector (C) of pass transistors Q404 through Q406. These transistors are wired in series to form a single high-voltage pass transistor network. At the nominal line voltage, the voltage at the collector (C) of transistor Q204 is approximately +300 V. The +300 and the −2000 V add together to produce the −1700 V that is supplied to the cathode of the CRT. Resistor R427 provides cathode current limiting. A voltage divider string, consisting of R423 through R426, supply a reduced voltage for FOCUS control R. A current-summing junction at the end of the divider adds to the high-voltage current, at approximately 340 μA, to a reference current supplied by the +15-V supply through control R419, and couples it to operational amplifier IC401. When the high voltage is at the correct level of −1700 V, control R419 is adjusted so the reference current is equal and opposite to the high-voltage current.

If the high voltage falls below −1700 V, the current difference at IC401 will cause it to increase the base drive to the pass transistors, Q404-Q406, and thus decrease the collector voltage until the high voltage again equals −1700 V. This will again make the high-voltage current equal to the reference current. The reverse will occur when the high voltage exceeds −1700 V.

The frequency response of this circuit is high enough so it also operates as a filter to remove 120-Hz ripple from the high voltage. Upper frequency response is limited by the capacitor C414 to suppress any possible high-frequency oscillation in U401. Diode D406 protects the pass transistor from a reverse output from U401. Resistors R411 through R413 ensure equal voltage distribution between Q404, Q405, and Q406.

CRT Blanking. The CRT blanking circuit is used to control the electron beam in the CRT. This includes blanking the CRT during retrace and hold-off, as well as adjusting the intensity of the beam. To understand this circuit, you must keep three things in mind:

- The CRT is blanked when the control grid is 68 V more negative than the cathode.
- As the 68-V difference between the grid and cathode is reduced, the CRT is unblanked, and the beam intensity is increased.
- Because the cathode of the CRT is at −1700 V, the grid must vary between −1700 and −1768 V. Therefore, the blanking circuit must be completely isolated from the other oscilloscope circuits. *Caution:* When you measure voltages in the blanking circuit, keep in mind that "circuit common" is 1700 V below oscilloscope ground.

Diodes D402 and ZD401 through ZD403, with capacitors C401 and C402, and resistor R401 make up a −68-V regulated power supply. The positive end of the supply is tied to the −1700-V supply. The −68-V supply powers a simple flip-flop, transistors Q402/Q403. This flip-flop is toggled through high-voltage capacitors C403 and C406 by blanking control IC203A. Because of the capacitor coupling, normal ground-referenced logic levels can be used for control.

Assume the flip-flop is toggled so Q402 is off and Q403 is on; this is the unblanked CRT condition. The cathode is near −1700 V and the other end of control R2 is at −1768 V, or −1700 V plus −68 V. As INTENSITY control R2 is turned clockwise, the beam intensity will increase at the same time the grid voltage approaches that on the cathode.

A blanking signal from IC203A, in a low-to-high logic transition at C406 and a high-to-low logic transition at C403, will toggle the flip-flop turning on Q402 and turning off Q403. Now both ends of control R2 are at −1768 V and the CRT will blank. The reverse will occur when an unblanked signal from IC203A is sensed. While resistors R404 and R405 hold the flip-flop in a stable state after each toggle, capacitors C404 and C405 speed up the switching cycle. The RC network R407/R408/C407 shapes the blanking signal. Resistor R408 also isolates the circuit.

Calibrator

An oscillator circuit, consisting of IC208A, IC208B and their associated circuits, generates an output of approximately 1000 Hz. This is coupled to the base (B) of transistor switch Q217. When Q217 is turned off, a precise 1-V level is connected to the output. Voltage divider resistors R506 through R508 set the output level. When Q217 is turned on, it grounds the output.

9
Understanding Digital Circuits

Modern electronics devices such as calculators and microcomputers make use of digital electronic circuits. These circuits are different from conventional analog electronic circuits in that only two voltage levels are needed to convey all of the necessary information.

Analog signals consist of infinitely variable voltages such as those observed in a sine wave. Digital signals, on the other hand, have only two distinct voltage levels, called HIGH and LOW. Other terms sometimes used for HIGH and LOW are ON and OFF, 1 and 0, and YES and NO. See Table 9-1.

HIGH	LOW
ON	OFF
1	0
YES	NO

Table 9-1. Digital Electronic States.

NUMBER SYSTEMS

To understand how digital electronic circuits function, it is necessary to understand the different number systems that are in use.

Decimal Number System

The number system that we use in everyday life is called the decimal number system. It is deemed so because it is based on the number 10. The ten numbers used are 0 through 9. To express a larger number we would add another digit to the left of the first. Figure 9-1.

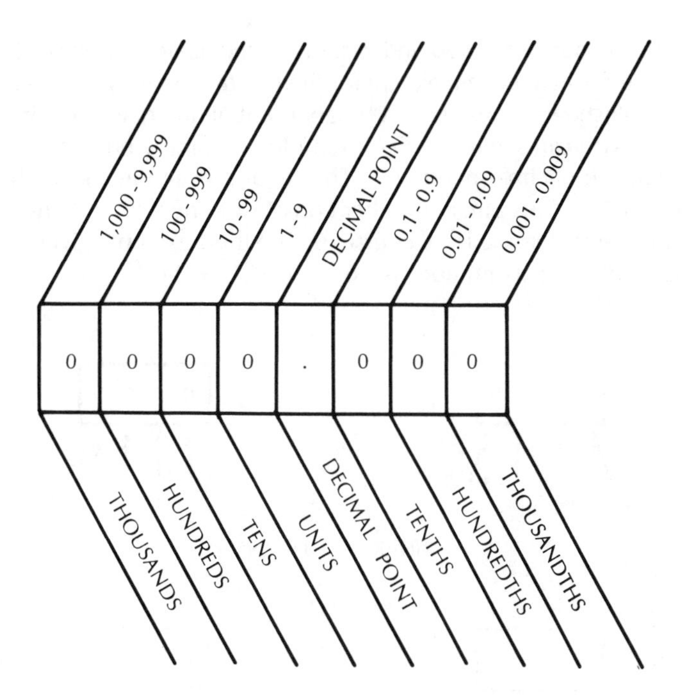

Fig. 9-1. The digital number system.

Binary Number System

The binary number system uses only two digits to express all possible numbers. These two digits are 0 and 1. They correspond to the digital logic states HIGH and LOW, ON and OFF, and YES and NO. See Table 9-2. It is not diffi-

Decimal	Binary	Decimal	Binary
0	00000	11	01011
1	00001	12	01100
2	00010	13	01101
3	00011	14	01110
4	00100	15	01111
5	00101	16	10000
6	00110	17	10001
7	00111	18	10010
8	01000	19	10011
9	01001	20	10100
10	01010	(and so on)	

Table 9-2. Comparison of Binary and Decimal Numbers.

cult to read binary numbers if you understand that the binary system works exactly the same as the decimal system except for the fact that only two digits are being used. Figure 9-2 shows how to read a binary number and determine its decimal equivalent. As you can see in the first example, the binary number 0000 0100 is the same as the decimal number 4. The second example shows the binary number 0000 0110. This binary number consists of one "4" and one "2". The decimal equivalent of 4 + 2 is, of course, 6. The third example gives the binary number 0001 0101. This number is 16 + 4 + 1 = 21.

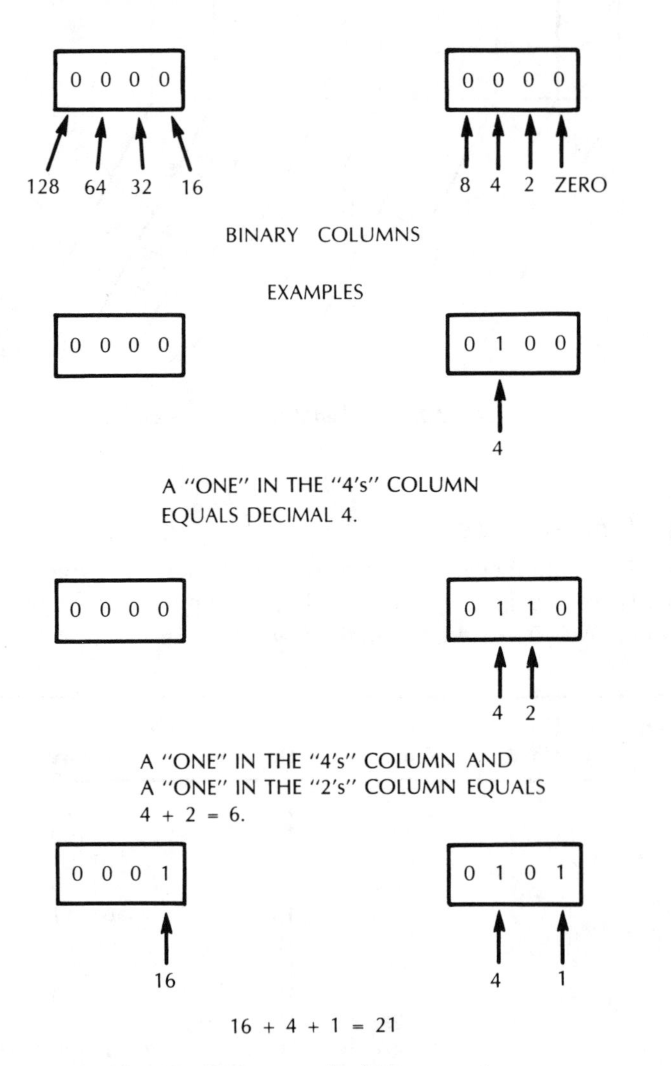

Fig. 9-2. How to read a binary number.

When writing binary numbers they should be broken down into groups of four. The binary number 0001 0101 is exactly the same as 00010101. Very large binary numbers would be extremely difficult to read if all of the ones and zeros were run together.

BOOLEAN ALGEBRA

To work with digital electronic circuits, it is necessary to have an understanding of the three basic functions in Boolean algebra. These three functions are AND, OR, and NOT.

The AND Function

In Boolean algebra, the output from an AND function will be a 1 (HIGH, YES, ON) if—and only if—both of the inputs are also a 1. If either input is a 0, then the output will be a 0. In other words, the output is HIGH if both A *and* B are HIGH. See Table 9-3.

Table 9-3. Truth Table for the AND Function.

A	B	A • B
0	0	0
0	1	0
1	0	0
1	1	1

Note: A•B is read "A *and* B." This is different than non-Boolean algebra where the dot would mean "*multiply* by."

The OR Function

The output from an OR function will be a 1 (HIGH, YES, ON) if either one of the two inputs is a 1. If both inputs are 0 then the output will also be a zero. In other words, the output will be HIGH if either A *or* B *or* both are high. See Table 9-4.

Table 9-4. Truth Table for the OR Function.

A	B	A + B
0	0	0
0	1	1
1	0	1
1	1	1

Note: A + B is read "A *or* B." This is different than non-Boolean algebra where the + would mean "*add*."

The NOT Function

The NOT function is the easiest function to remember. It simply reverses (inverts) the logic condition of the input. The NOT function has only one input. If the input is a 1, then the output will be a 0. If the input is a 0, then the output will be a 1. See Table 9-5.

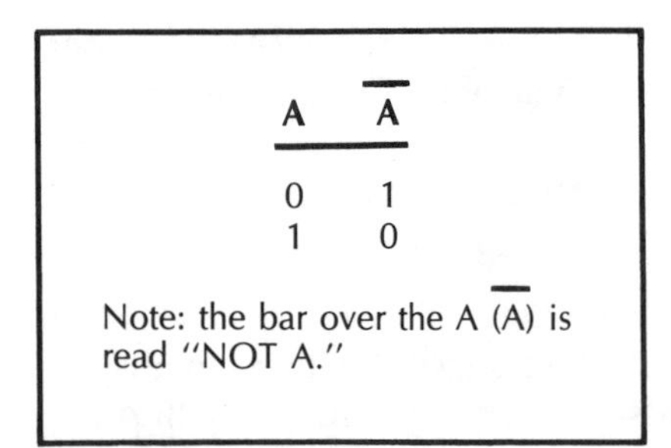

A	$\overline{A}$
0	1
1	0

Note: the bar over the A ($\overline{A}$) is read "NOT A."

Table 9-5. Truth Table for the NOT Function.

DIGITAL GATES

The basic component of a digital electronic circuit is the *gate*. These gates are components that can accept one or more inputs and produce an output according to the rules of Boolean algebra. Digital gates could be implemented with simple switches but such units would be extremely bulky! They could also be implemented with vacuum tubes but, again, they would be bulky and use a lot of power.

Digital ICs

Digital integrated circuits contain anywhere from a few simple gates to many thousands of gates. An SSI (small-scale integration) integrated circuit like the 4069 contains six gates. See Fig. 9-3.

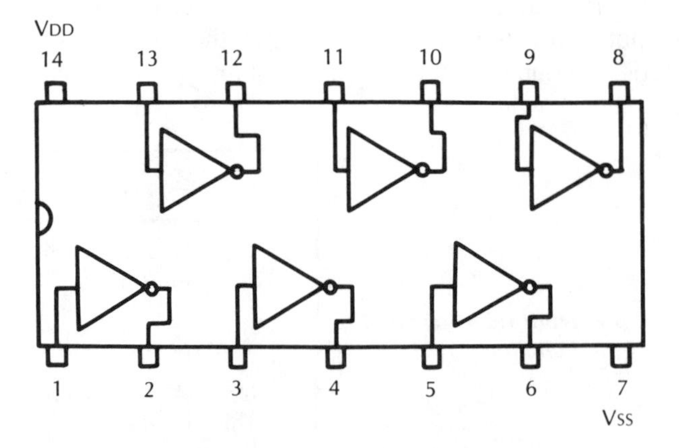

Fig. 9-3. The 4069 IC contains six NOT gates.

The AND Gate

The AND Gate will give a HIGH output only if all of its inputs are HIGH. Figure 9-4 shows the schematic symbol for a two-input AND gate. Many AND gates have three, four, or even more inputs. They work the same way as a two-input AND gate.

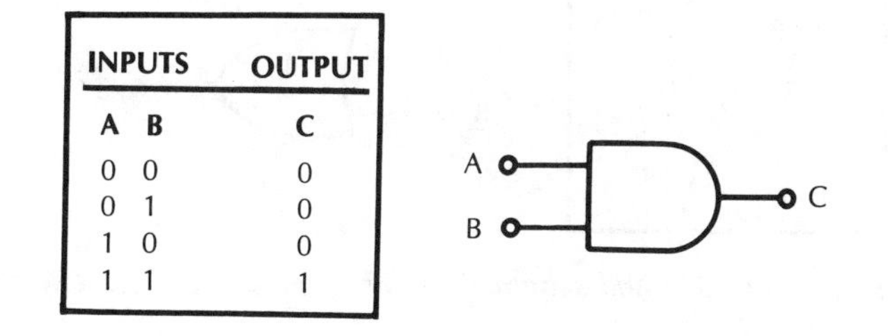

INPUTS		OUTPUT
A	B	C
0	0	0
0	1	0
1	0	0
1	1	1

Fig. 9-4. Truth table and schematic symbol for a two-input AND gate.

The AND function is written A•B = C or AB = C. The dot between the A and B does not mean multiplication; it represents AND Boolean algebra.

The AND gate can be thought of as two or more switches in series. See Fig. 9-5. Both switches A and B must be closed (1) for the lamp to light. If either or both of the switches are open (0), then the lamp will not be lit (0).

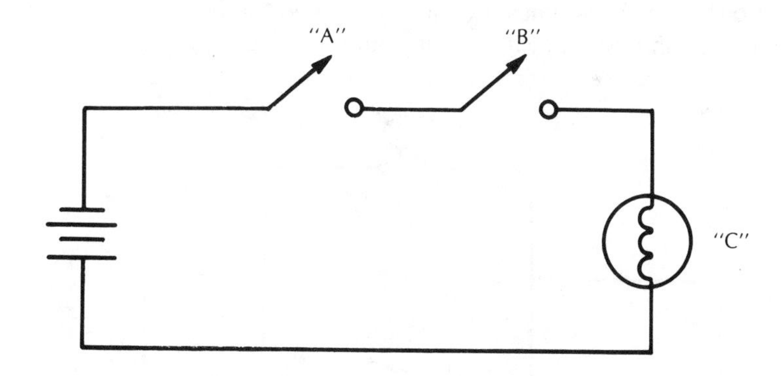

Fig. 9-5. A two-input AND circuit.

The OR Gate

The OR gate will give a HIGH output if any one of its inputs are HIGH. Figure 9-6 shows the schematic symbol for a two-input OR gate.

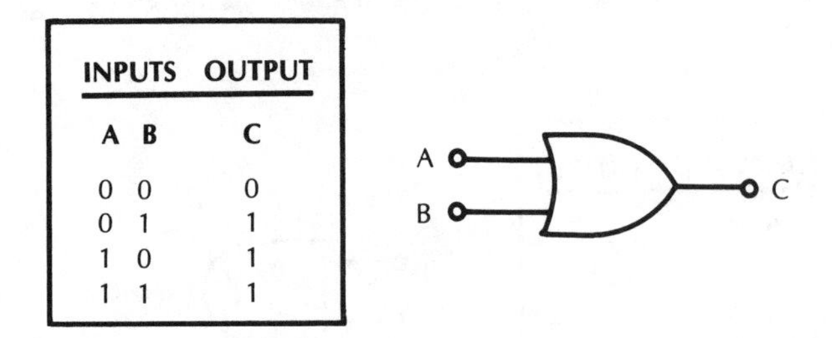

INPUTS		OUTPUT
A	B	C
0	0	0
0	1	1
1	0	1
1	1	1

Fig. 9-6. Truth table and schematic symbol for a two-input OR gate.

The OR function is written A + B = C. The (+) sign does not mean to add; it means OR in Boolean algebra. To begin working with digital logic, you should memorize the AND and OR gate logic functions:

A•B = C If A is HIGH *AND* B is HIGH, then C is HIGH.

A+B = C If either A *OR* B is HIGH, then C is HIGH.

The OR gate may be thought of as two switches in parallel. See Fig. 9-7. If either one or both switches are closed (1) then the lamp will light (1). The only way the lamp will not light is if both switches are open.

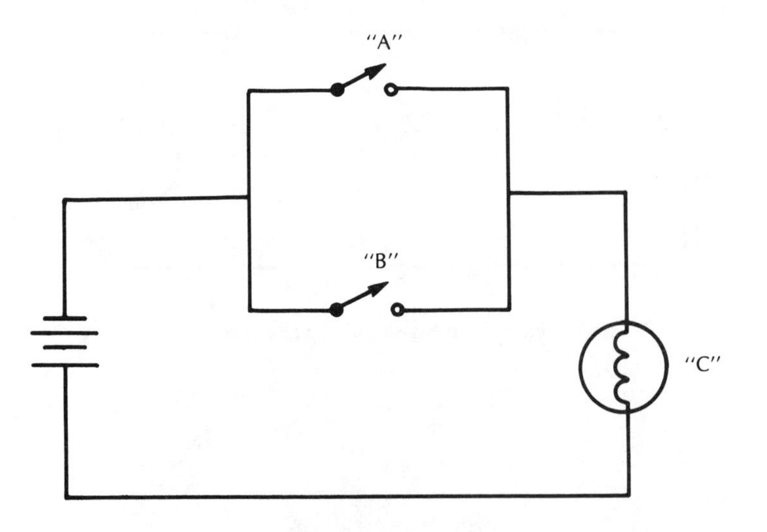

Fig. 9-7. A two-input OR circuit.

The NAND Gate

The NAND gate is the opposite of the AND gate. The NAND gate will give a LOW output if all of its inputs are HIGH. If any of its inputs are LOW, then it will give a HIGH output. Figure 9-8 shows the schematic symbol and truth table for a NAND gate. The small circle at the right of the symbol is called an *inverter*. This symbol means that this is *N*ot an AND gate (NAND).

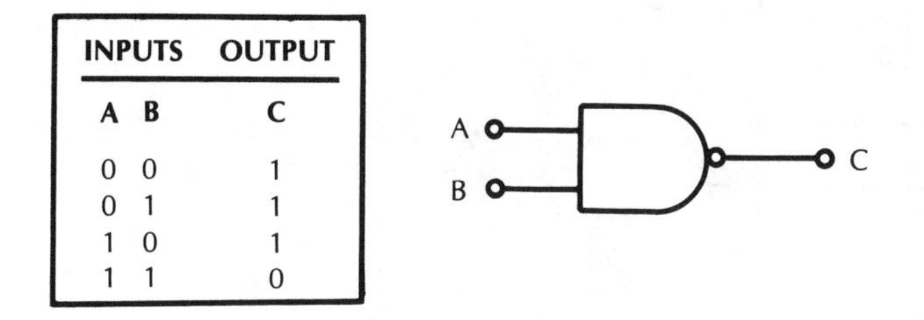

INPUTS A	INPUTS B	OUTPUT C
0	0	1
0	1	1
1	0	1
1	1	0

Fig. 9-8. Truth table and schematic symbol for a two-input NAND gate.

The NAND function is written $A \bullet B = \overline{C}$. This mean "A and B equals NOT C." The overscore above the C indicates an inverted condition.

The NOR Gate

The NOR gate is the opposite of the OR gate. The NOR gate will give a HIGH output only if all of its inputs are LOW. If any of its inputs are HIGH it will give a LOW output. Figure 9-9 shows the schematic symbol and truth table for a two-input NOR gate. Notice the small inverter circle on the right side of the gate. This means that this is an inverted OR gate or a *N*ot OR (NOR).

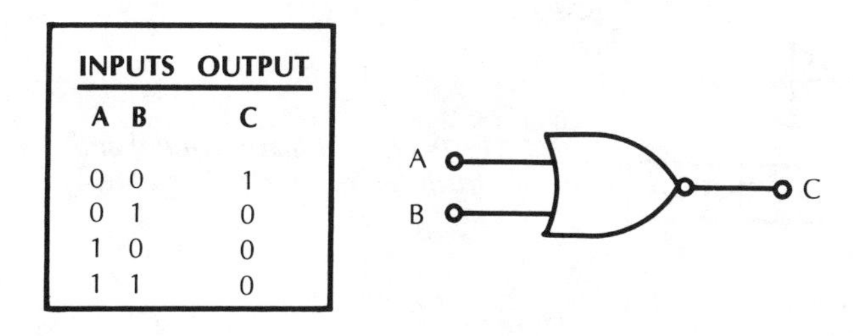

INPUTS A	INPUTS B	OUTPUT C
0	0	1
0	1	0
1	0	0
1	1	0

Fig. 9-9. Truth table and schematic symbol for a two-input NOR gate.

The NOR gate function is written $A + B = \overline{C}$. This means that A *OR B* equal NOT C. The overscore above the C indicates an inverted condition.

The Exclusive-OR Gate

The Exclusive-OR Gate (XOR) will give a HIGH output if one or more of its inputs is HIGH, but the inputs must have different logic states. Figure 9-10 shows the schematic symbol and truth table for the XOR gate.

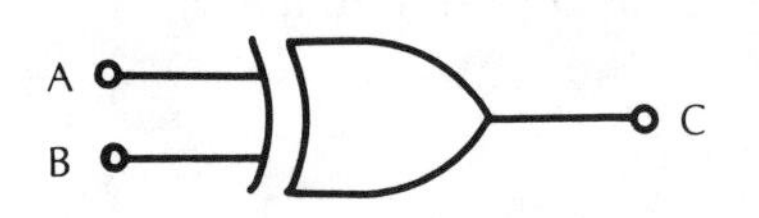

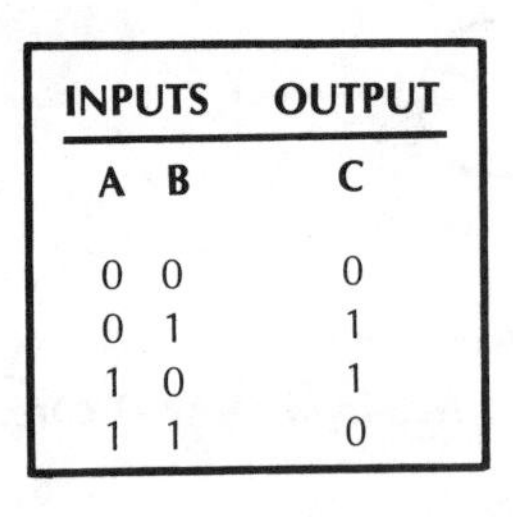

INPUTS A	INPUTS B	OUTPUT C
0	0	0
0	1	1
1	0	1
1	1	0

Fig. 9-10. Exclusive-OR gate schematic symbol and truth table.

The XOR gate has a curved bar at the input side of the symbol. The XOR function is written $A \oplus B = C$. Notice the circle around the + sign. This circle indicates that it is an exclusive-OR function rather than an OR function.

The Buffer Gate

The buffer gate is really very simple. It has one input and one output. The buffer is used as an amplifier. It does not change the digital logic. The schematic symbol and truth table is shown in Fig. 9-11.

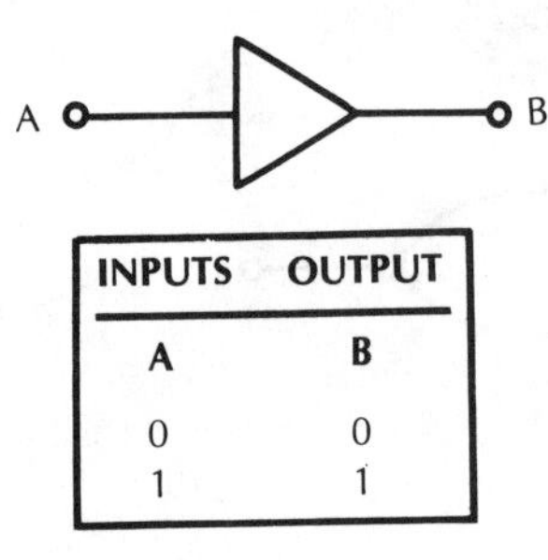

INPUTS A	OUTPUT B
0	0
1	1

Fig. 9-11. Schematic symbol and truth table for a buffer gate.

The NOT Gate

The NOT gate is very similar to the buffer except that it inverts the input. Figure 9-12 shows the schematic symbol and truth table for a NOT gate.

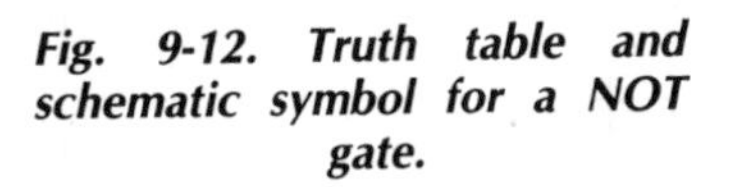

Fig. 9-12. Truth table and schematic symbol for a NOT gate.

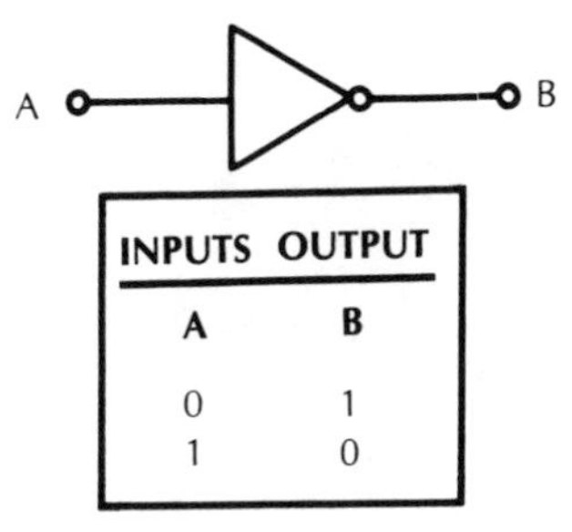

INPUTS	OUTPUT
A	B
0	1
1	0

READING DIGITAL SCHEMATICS

A knowledge of the basic digital gates will enable you to understand digital circuit diagrams. For example, referring back to Fig. 9-3, you can see that this 14-pin dual in-line package (DIP) integrated circuit contains six NOT gates. If you were to input a HIGH at pin 1, you would be able to read a LOW at pin 2. If you were to input this LOW into pin 3 you would then read a HIGH at pin 4. See Figs. 9-13 and 9-14 for examples of digital schematics.

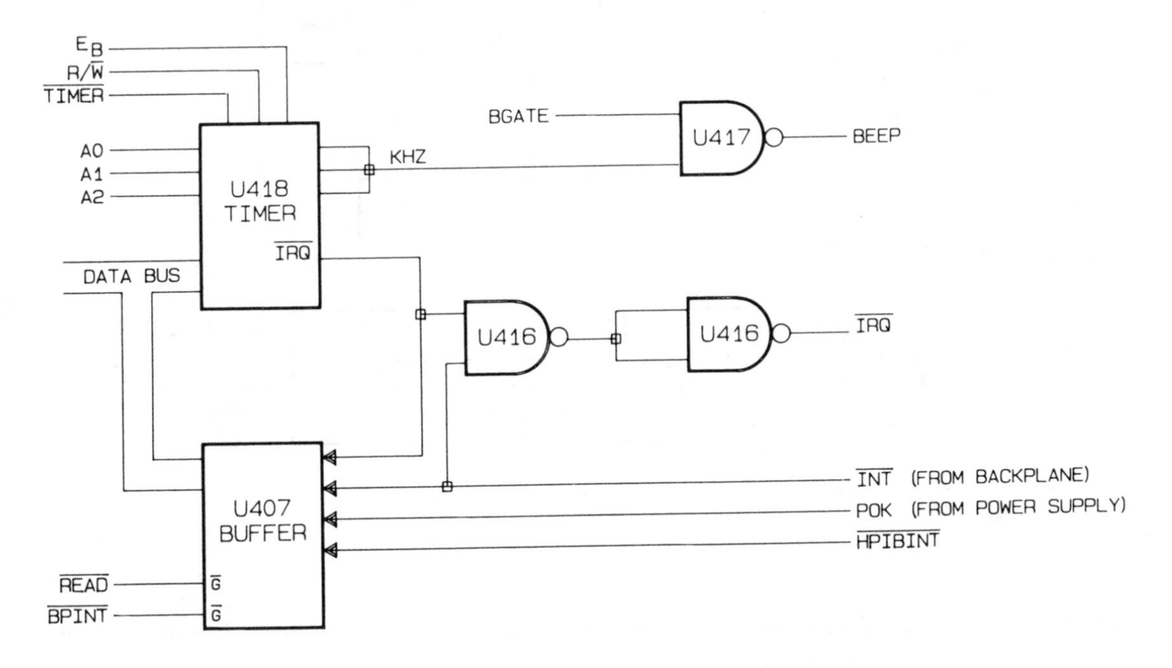

Fig. 9-13. A logic schematic (courtesy of Hewlett Packard).

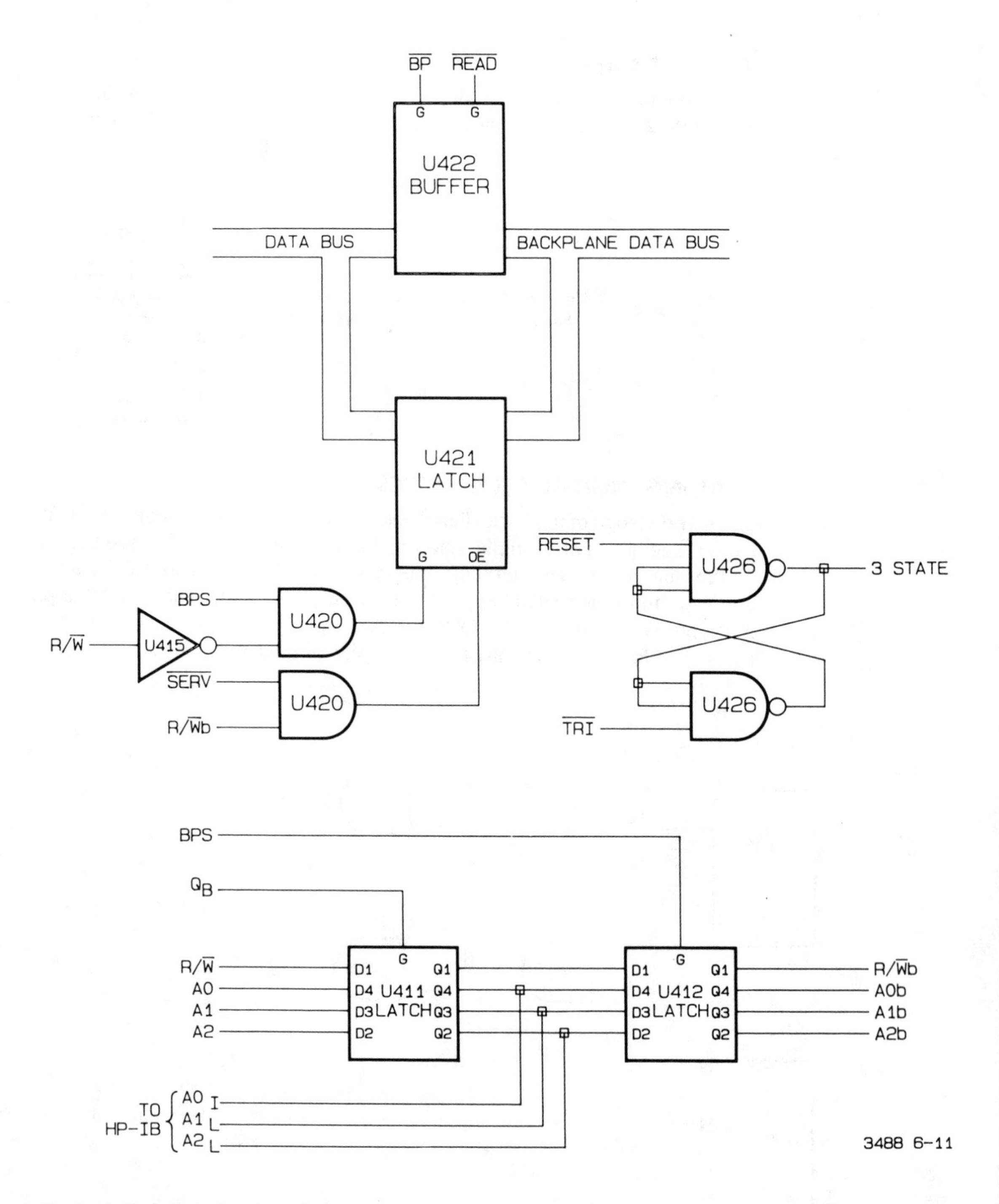

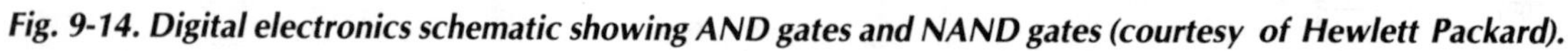
Fig. 9-14. Digital electronics schematic showing AND gates and NAND gates (courtesy of Hewlett Packard).

MEASURING LOGIC LEVELS

The logic levels in digital circuits are measured with a test instrument called a logic probe. When the logic probe's test probe is touched to a test point, an LED will light up to show the logic level (HIGH or LOW). A good logic probe will also indicate pulses and can be used for both TTL and CMOS circuits. Figure 9-15 shows a logic probe that can do all of these things.

Fig. 9-15. A good logic probe is absolutely essential for experimenting with digital circuits (courtesy of Radio Shack, A Division of Tandy Corporation).

Appendix A

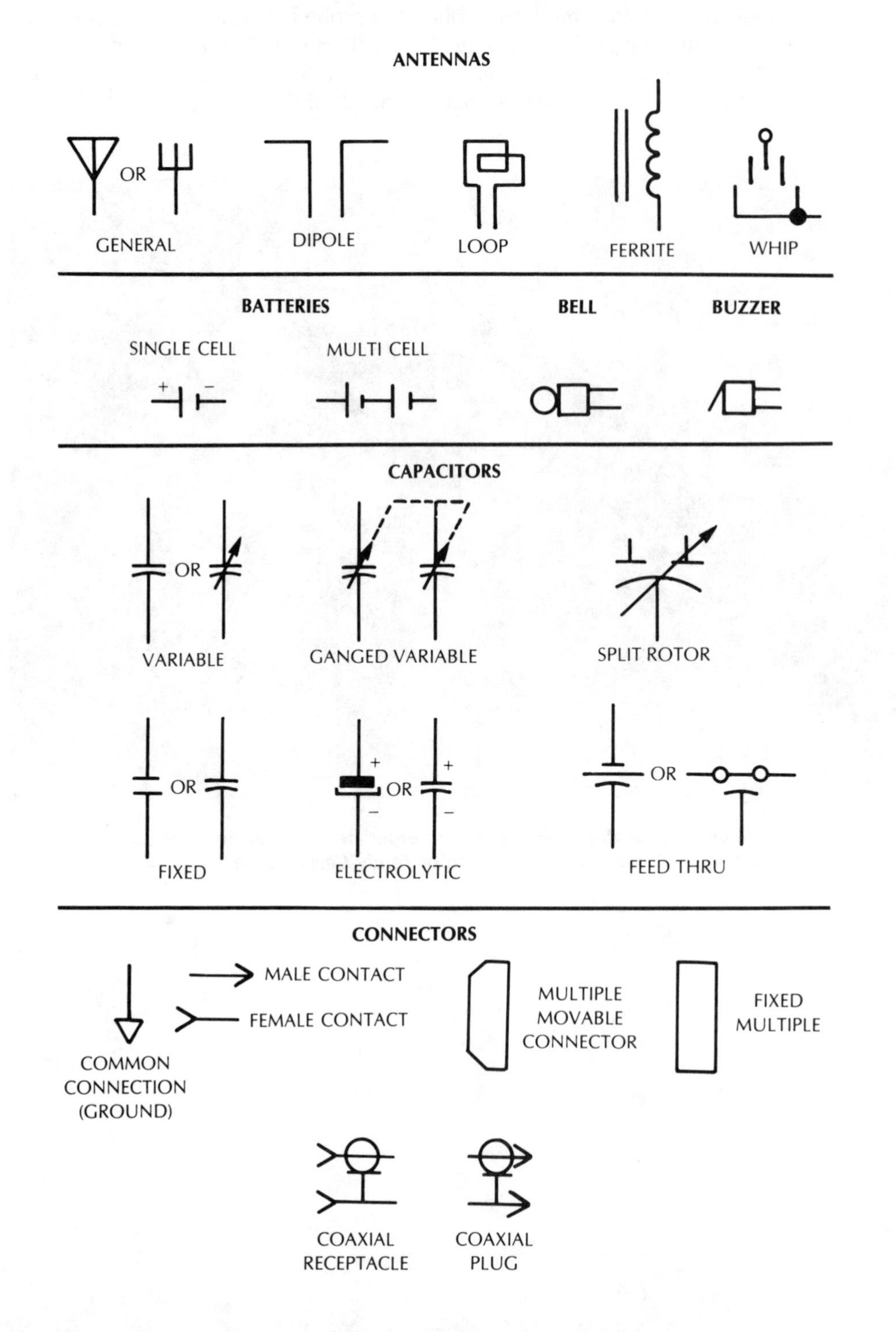
ANTENNAS
OR
GENERAL
DIPOLE
LOOP
FERRITE
WHIP
BATTERIES
SINGLE CELL
+
−
MULTI CELL
BELL
BUZZER
CAPACITORS
OR
VARIABLE
GANGED VARIABLE
SPLIT ROTOR
OR
FIXED
+
OR
−
+
−
ELECTROLYTIC
OR
FEED THRU
CONNECTORS
MALE CONTACT
FEMALE CONTACT
COMMON CONNECTION (GROUND)
MULTIPLE MOVABLE CONNECTOR
FIXED MULTIPLE
COAXIAL RECEPTACLE
COAXIAL PLUG

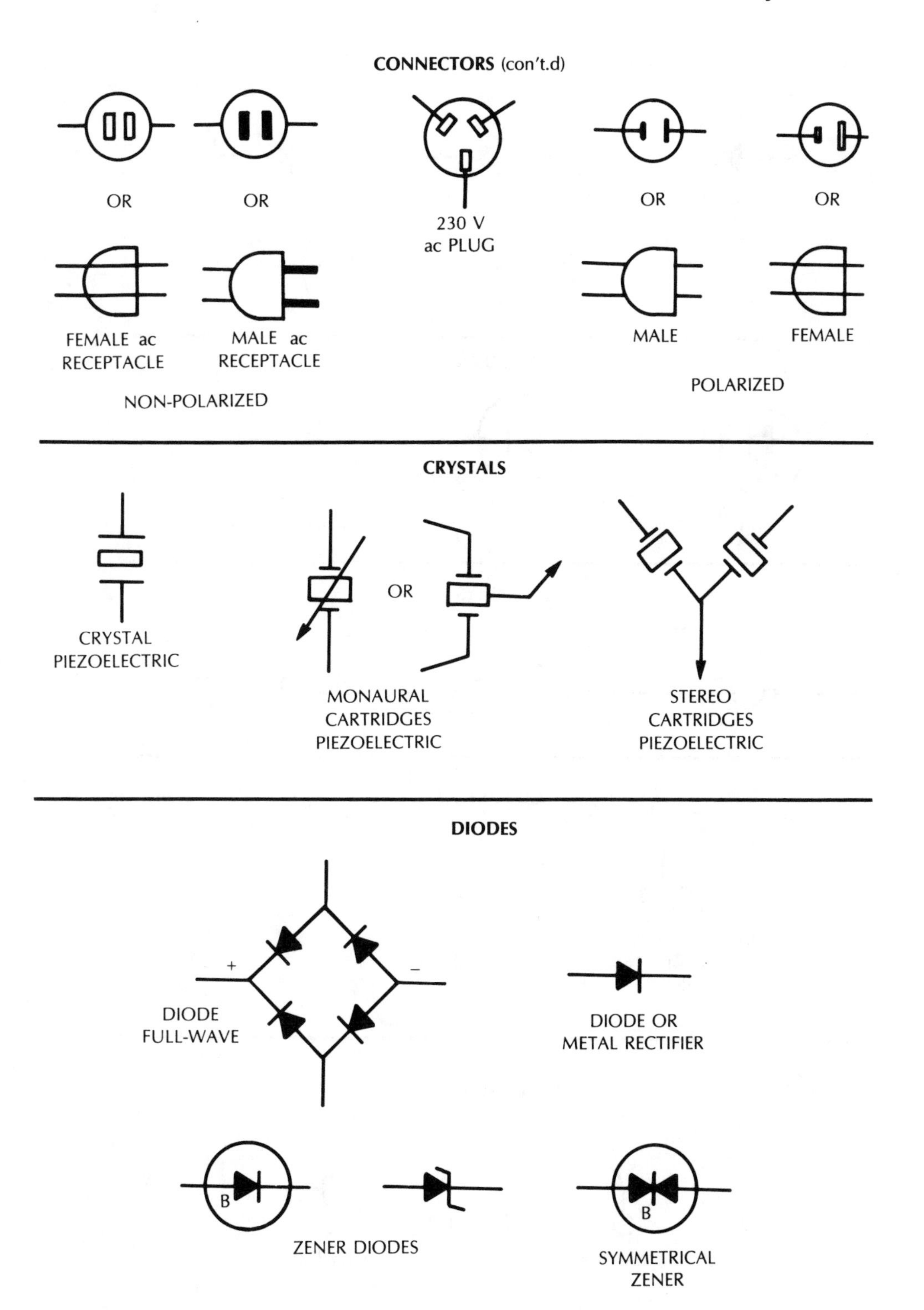
CONNECTORS (con't.d)
OR
OR
230 V
ac PLUG
OR
OR
FEMALE ac
RECEPTACLE
MALE ac
RECEPTACLE
NON-POLARIZED
MALE
FEMALE
POLARIZED
CRYSTALS
CRYSTAL
PIEZOELECTRIC
OR
MONAURAL
CARTRIDGES
PIEZOELECTRIC
STEREO
CARTRIDGES
PIEZOELECTRIC
DIODES
+
–
DIODE
FULL-WAVE
DIODE OR
METAL RECTIFIER
B
ZENER DIODES
B
SYMMETRICAL
ZENER

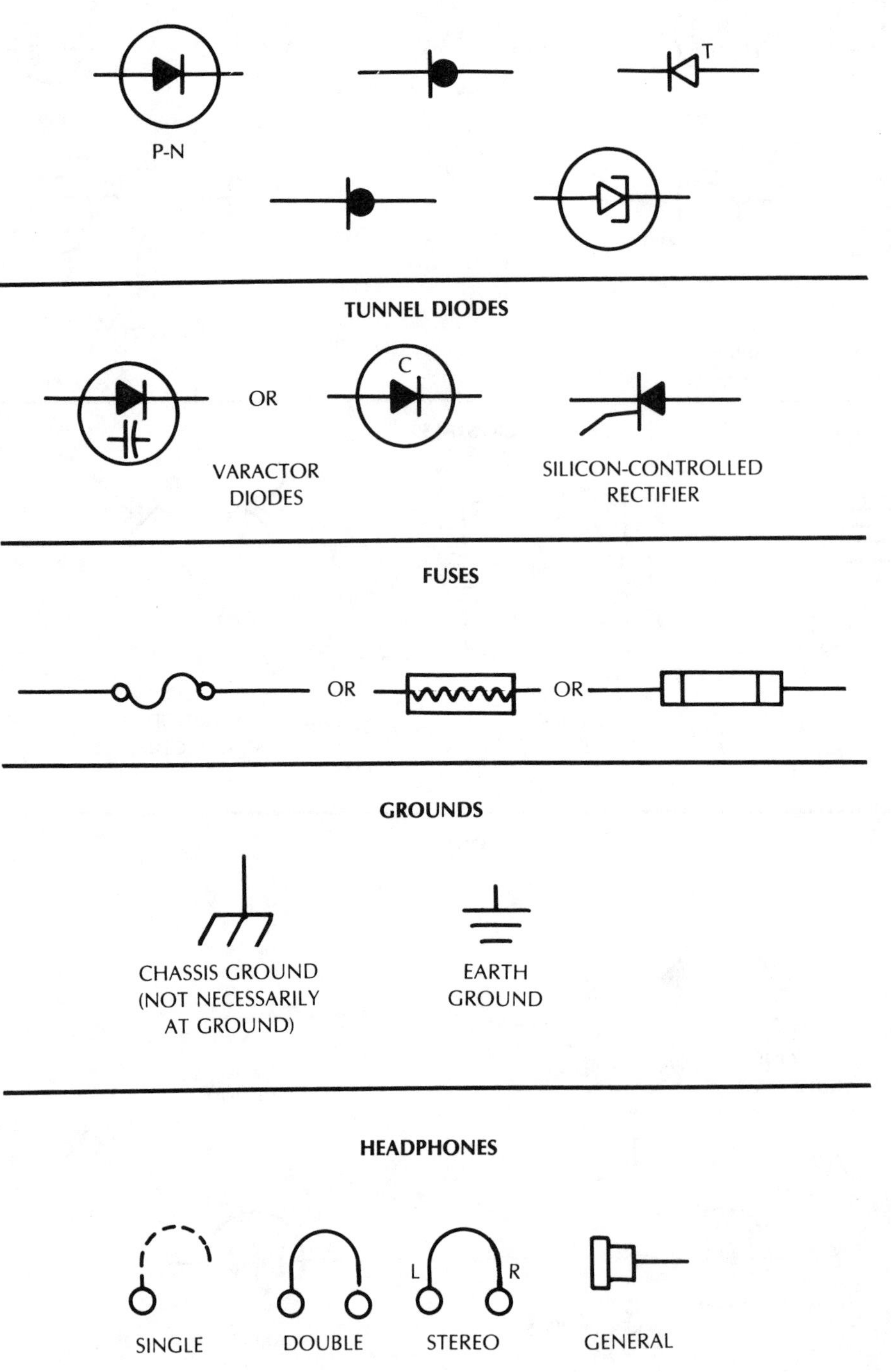
T
P-N
TUNNEL DIODES
C
OR
VARACTOR
DIODES
SILICON-CONTROLLED
RECTIFIER
FUSES
OR
OR
GROUNDS
CHASSIS GROUND
(NOT NECESSARILY
AT GROUND)
EARTH
GROUND
HEADPHONES
L
R
SINGLE
DOUBLE
STEREO
GENERAL

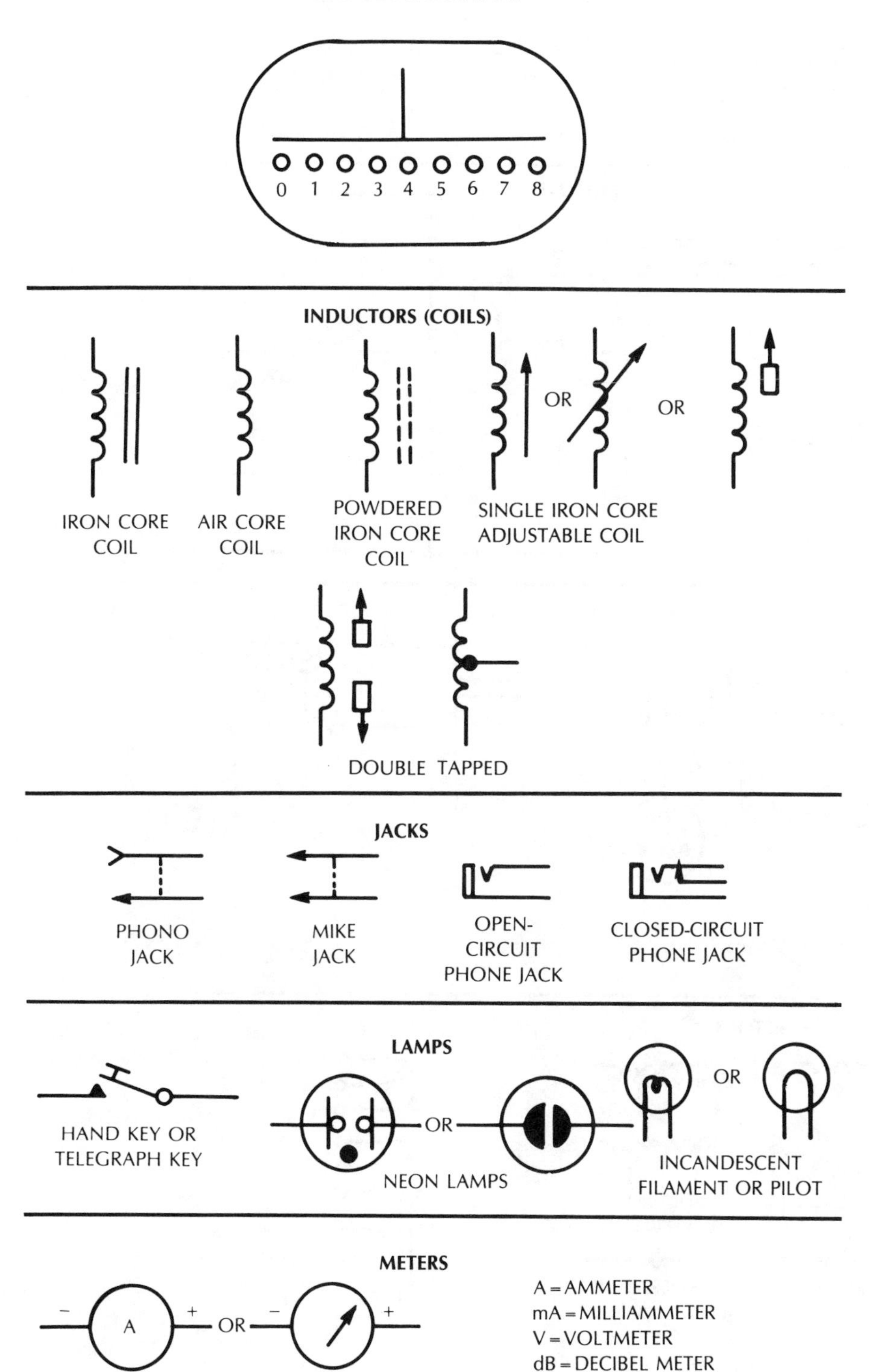
READOUT INDICATOR
0 1 2 3 4 5 6 7 8
INDUCTORS (COILS)
IRON CORE COIL
AIR CORE COIL
POWDERED IRON CORE COIL
OR
OR
SINGLE IRON CORE ADJUSTABLE COIL
DOUBLE TAPPED
JACKS
PHONO JACK
MIKE JACK
OPEN-CIRCUIT PHONE JACK
CLOSED-CIRCUIT PHONE JACK
LAMPS
HAND KEY OR TELEGRAPH KEY
OR
NEON LAMPS
OR
INCANDESCENT FILAMENT OR PILOT
METERS
A
OR
A = AMMETER
mA = MILLIAMMETER
V = VOLTMETER
dB = DECIBEL METER

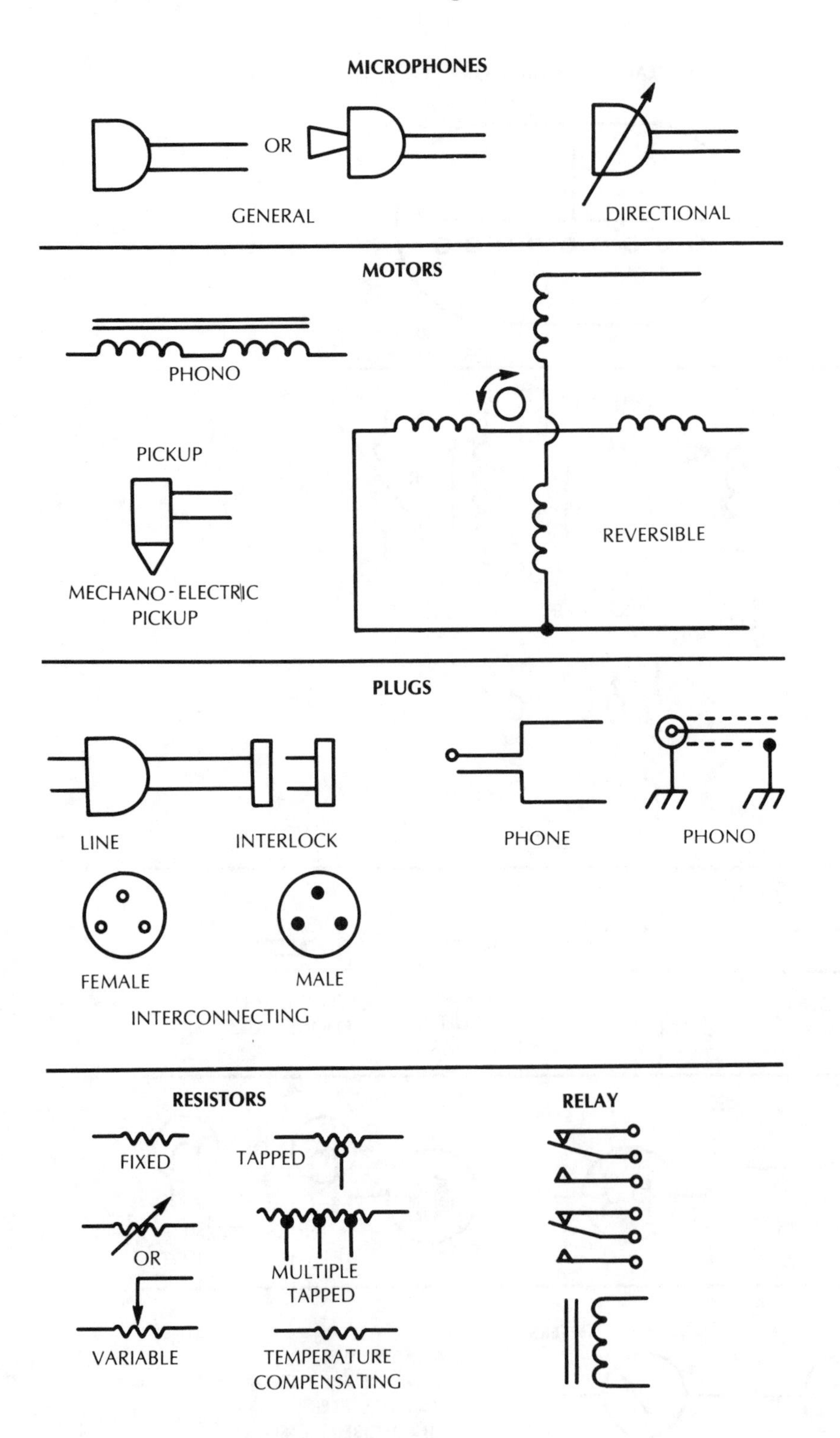
MICROPHONES
OR
GENERAL
DIRECTIONAL
MOTORS
PHONO
PICKUP
MECHANO-ELECTRIC
PICKUP
REVERSIBLE
PLUGS
LINE
INTERLOCK
PHONE
PHONO
FEMALE
MALE
INTERCONNECTING
RESISTORS
FIXED
TAPPED
OR
MULTIPLE
TAPPED
VARIABLE
TEMPERATURE
COMPENSATING
RELAY

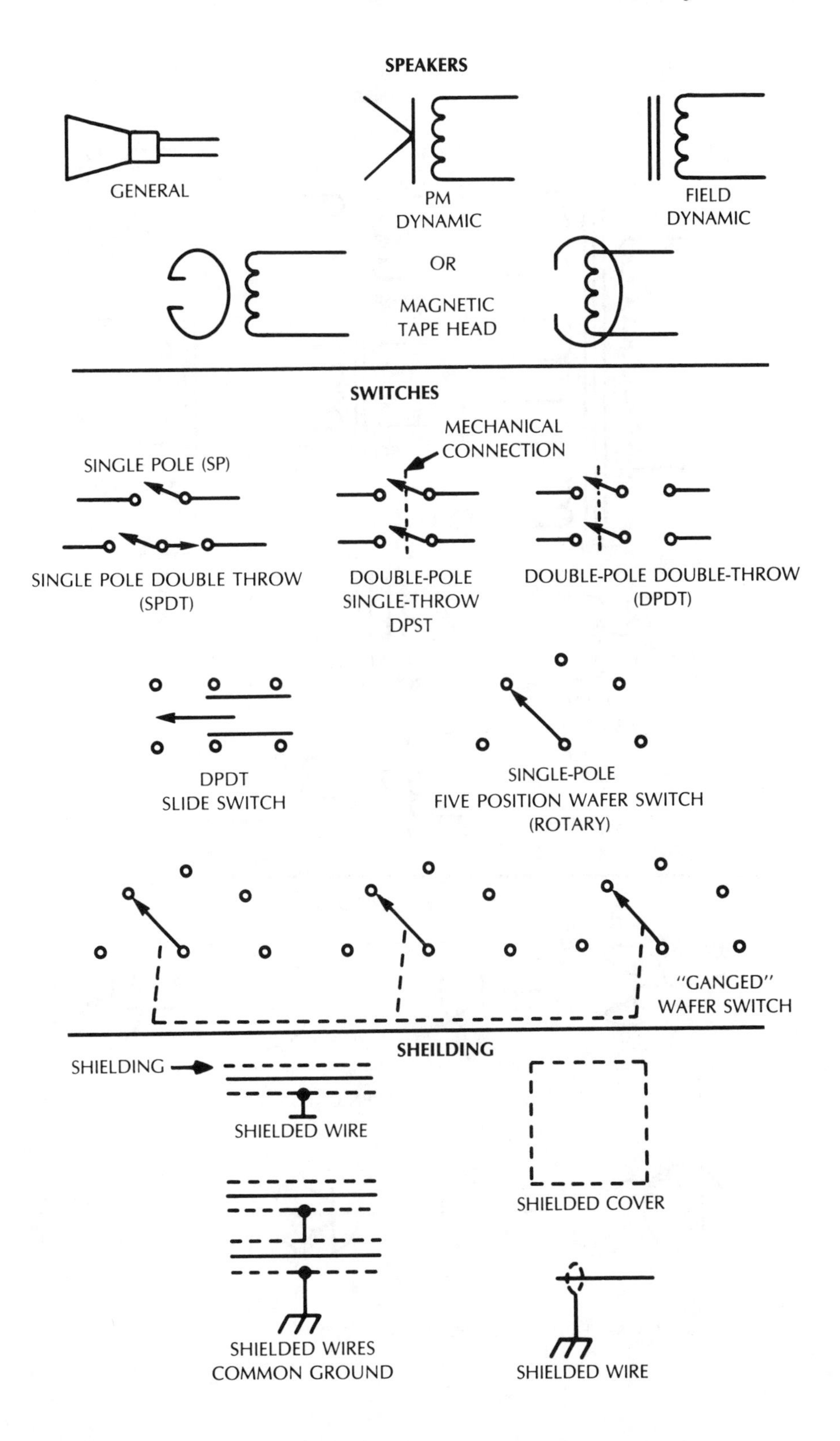
SPEAKERS
GENERAL
PM
DYNAMIC
FIELD
DYNAMIC
OR
MAGNETIC
TAPE HEAD
SWITCHES
MECHANICAL
CONNECTION
SINGLE POLE (SP)
SINGLE POLE DOUBLE THROW
(SPDT)
DOUBLE-POLE
SINGLE-THROW
DPST
DOUBLE-POLE DOUBLE-THROW
(DPDT)
DPDT
SLIDE SWITCH
SINGLE-POLE
FIVE POSITION WAFER SWITCH
(ROTARY)
"GANGED"
WAFER SWITCH
SHEILDING
SHIELDING
SHIELDED WIRE
SHIELDED COVER
SHIELDED WIRES
COMMON GROUND
SHIELDED WIRE

TRANSFORMERS

POWER

AIR CORE

POWERED
IRON CORE

IRON CORE

AUTO
TRANSFORMER

ADJUSTABLE
AIR CORE

ADJUSTABLE
IRON CORE

OR

LINK COUPLED

TRANSISTORS

NPN

PNP

PN UNIJUNCTION

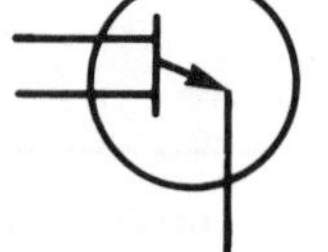

NP UNIJUNCTION

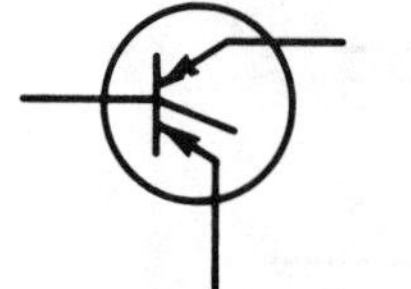

PNPN

NPN
(HOOK OR CONJUGATE-EMITTER CONNECTION)

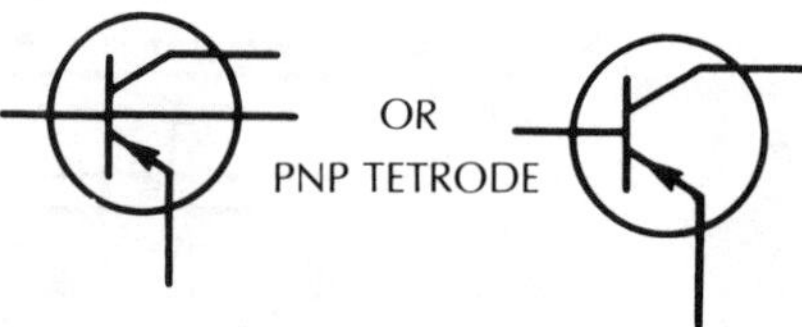

PNP TETRODE

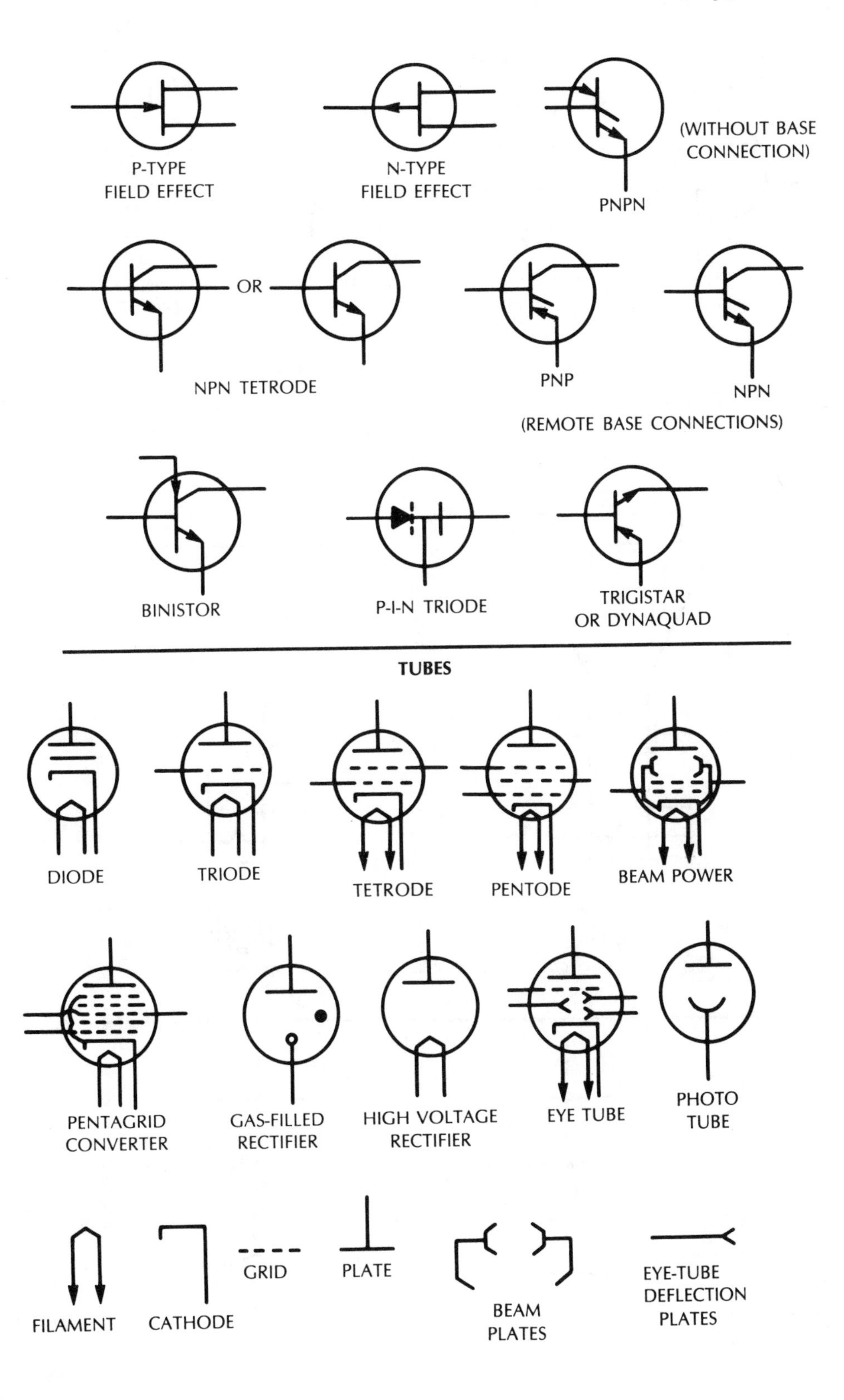
P-TYPE
FIELD EFFECT
N-TYPE
FIELD EFFECT
PNPN
(WITHOUT BASE
CONNECTION)
OR
NPN TETRODE
PNP
NPN
(REMOTE BASE CONNECTIONS)
BINISTOR
P-I-N TRIODE
TRIGISTAR
OR DYNAQUAD
TUBES
DIODE
TRIODE
TETRODE
PENTODE
BEAM POWER
PENTAGRID
CONVERTER
GAS-FILLED
RECTIFIER
HIGH VOLTAGE
RECTIFIER
EYE TUBE
PHOTO
TUBE
FILAMENT
CATHODE
GRID
PLATE
BEAM
PLATES
EYE-TUBE
DEFLECTION
PLATES

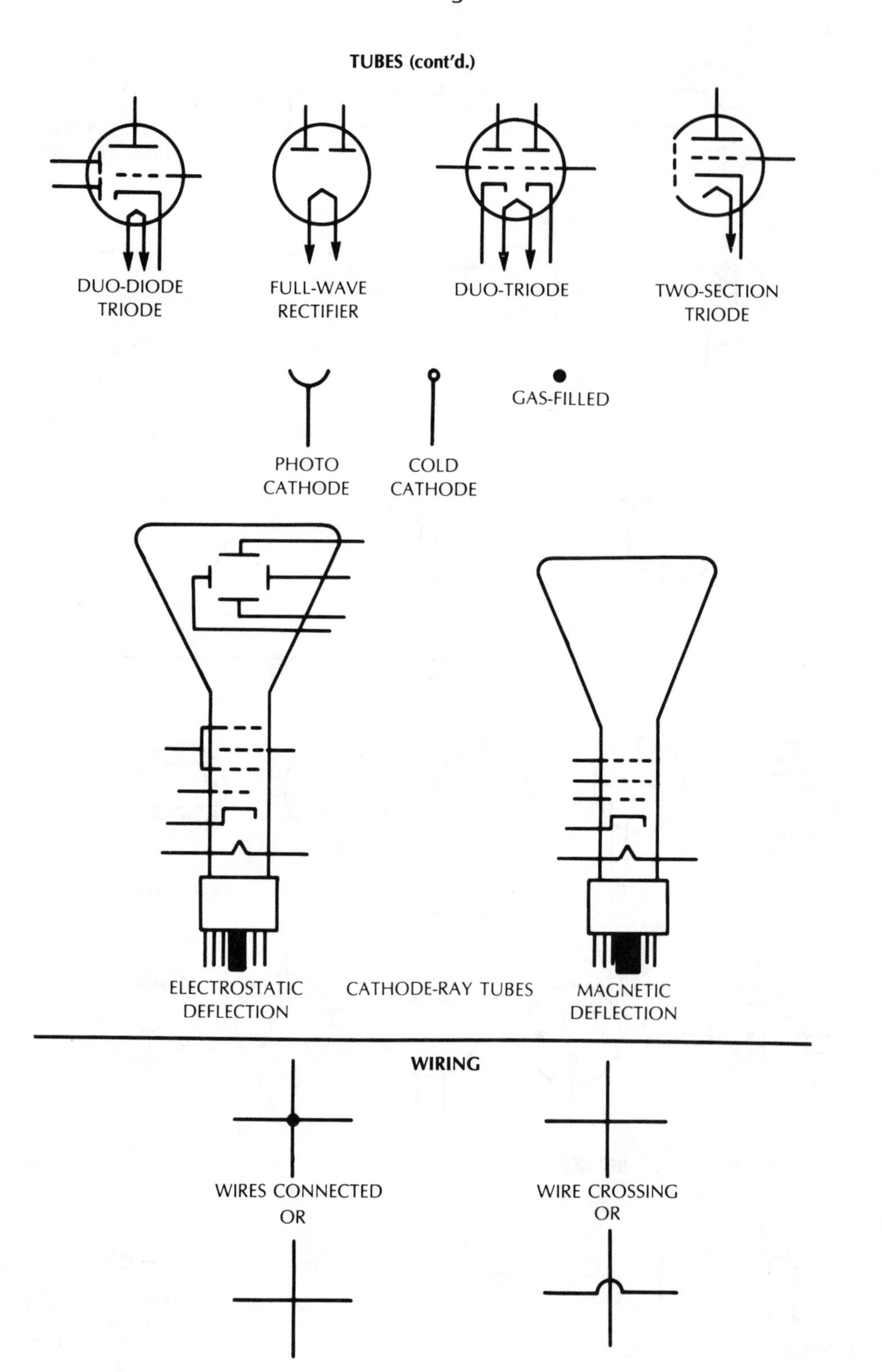
TUBES (cont'd.)
DUO-DIODE TRIODE
FULL-WAVE RECTIFIER
DUO-TRIODE
TWO-SECTION TRIODE
GAS-FILLED
PHOTO CATHODE
COLD CATHODE
ELECTROSTATIC DEFLECTION
CATHODE-RAY TUBES
MAGNETIC DEFLECTION
WIRING
WIRES CONNECTED
OR
WIRE CROSSING
OR

Appendix B

COLOR CODES FOR CAPACITORS

COLOR	DIGIT	MULTLIPLIER	TOLERANCE
BLACK	0	1	
BROWN	1	10	
RED	2	100	
ORANGE	3	1000	
YELLOW	4	10000	
GREEN	5		
BLUE	6		
VIOLET	7		
GRAY	8		
WHITE	9		
GOLD			5%
SILVER			10%
NO COLOR			20%

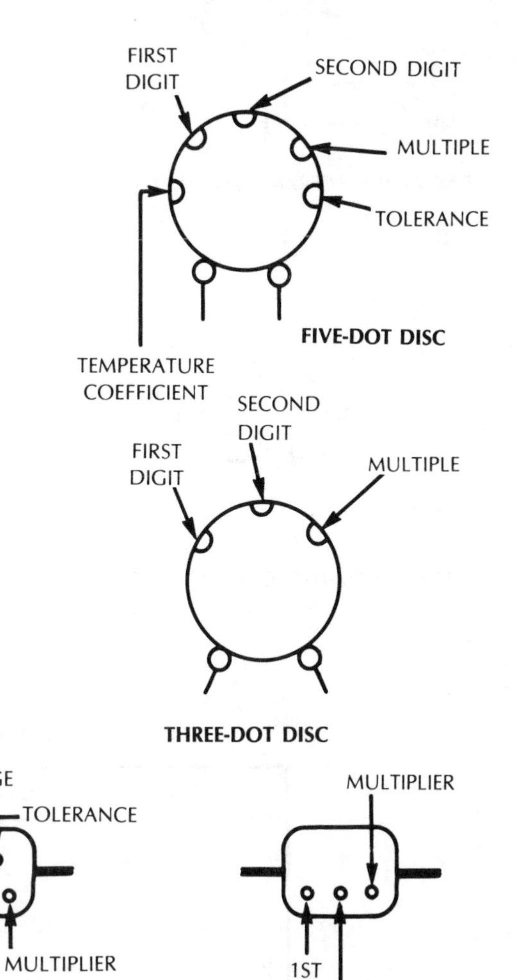

MOLDED CERAMIC

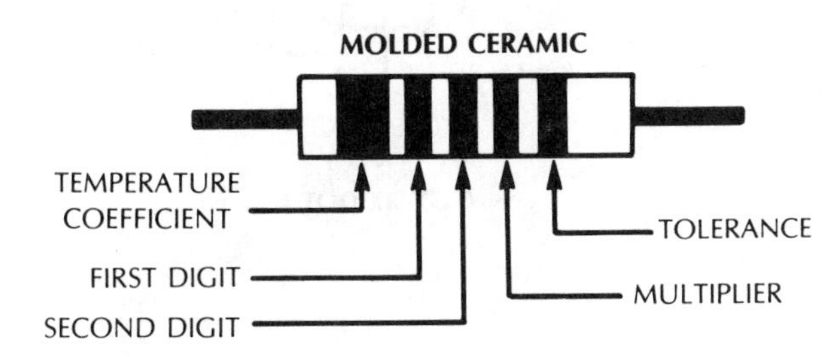

2ND DIGIT
MULTIPLIER
1ST DIGIT
TOLERANCE
VOLTAGE

VOLTAGE
TOLERANCE
1ST DIGIT
2ND
MULTIPLIER

MULTIPLIER
1ST
2ND DIGIT

MICA CAPACITOR
(500 WVDC AT ± 20%)

*NOTE
MICA CAPACITORS MAY BE CODED AS ABOVE IN OLD RMA CODE OR IN PRESENT EIA CODE. A WHITE DOT IS EIA, BLACK IS MILITARY AND SILVER IS AMERICAN WAR STANDARD. ANY OTHER COLOR FOR THE FIRST DOT IS THE OLD EIA CODING AND THE FIRST FOUR DOTS SHOW CAPACITOR VALUE.

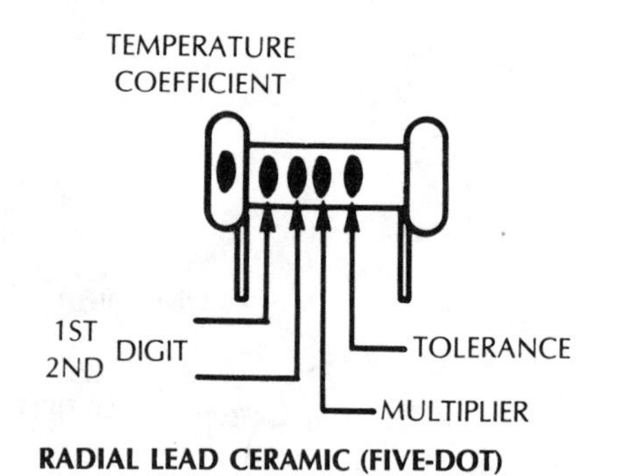

RADIAL LEAD CERAMIC (FIVE-DOT)

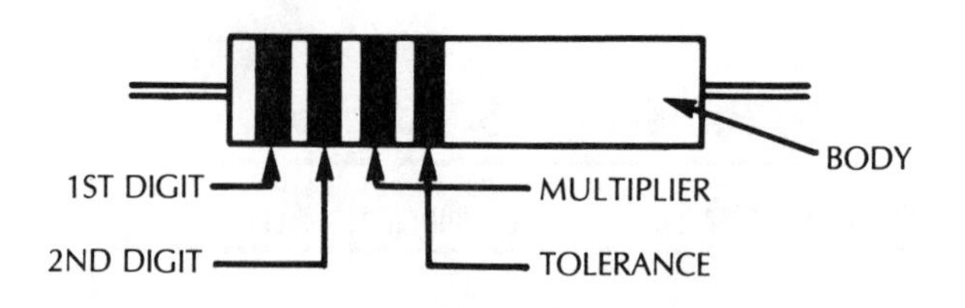

RESISTOR COLOR CODE

WHT, EIA
BLK, MIL
SILVER, AWS
1ST
2ND DIGITS
MULTIPLIER
CLASSIFICATION
TOLERANCE

PRESENT SIX-DOT CODE *SEE NOTE

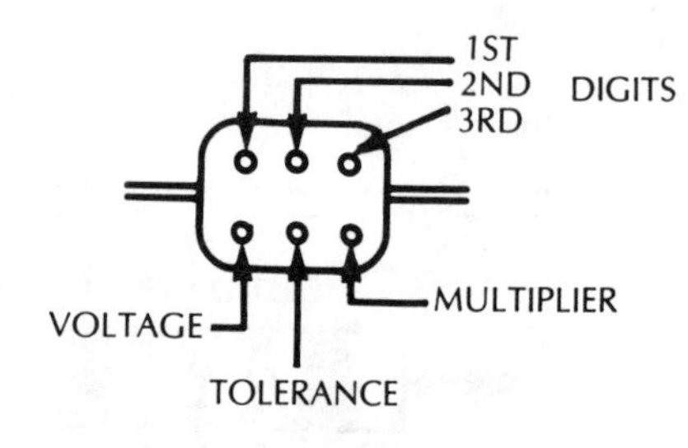

OLD SIX-DOT CODE

EIA COLOR CODE

COLOR	DIGIT	MULTIPLIER	TOLERANCE
BLACK	0		
BROWN	1	0	
RED	2	00	
ORANGE	3	000	
YELLOW	4	0000	
GREEN	5	00000	
BLUE	6	000000	
VIOLET	7		
GRAY	8		
WHITE	9		
GOLD		0.1	±5%
SILVER		0.01	±10%
NO BAND			±20%

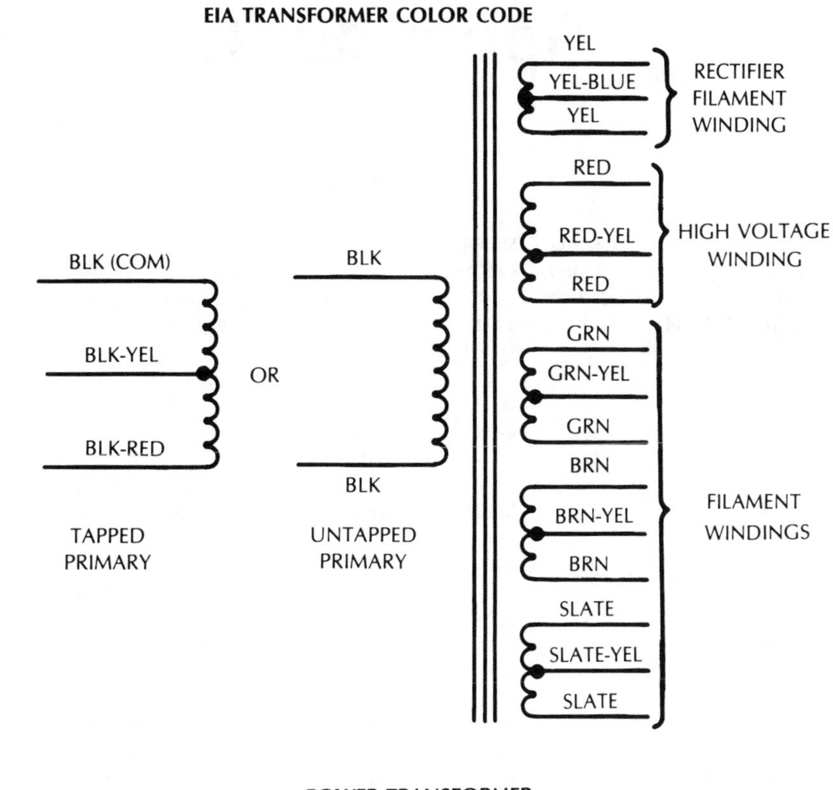

POWER TRANSFORMER

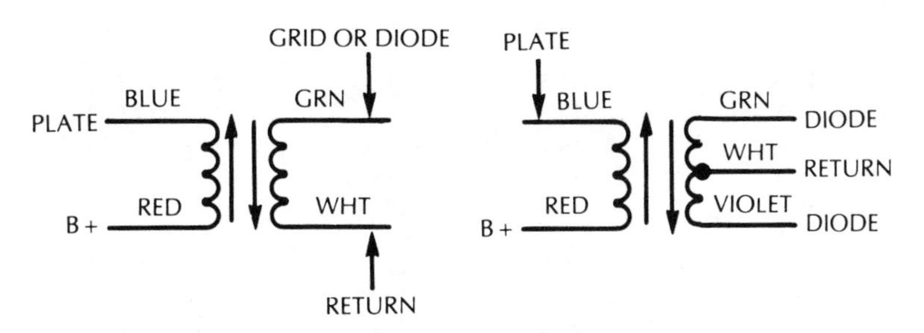

i-f TRANSFORMERS

AUDIO OUTPUT AND INTERSTAGE TRANSFORMERS

GRID OR
VOICE COIL
BLUE
GRN
PLATE
RED
BLK
B+
RETURN
PLATE
BLUE
GRN
VOICE COIL
RED
B+
BRN
BLK
PLATE
RETURN
PLATE
BLUE
GRN
GRID
BLK
RETURN
YEL
RED
B+
GRID

Appendix C

Wires Sizes

AWG B&S GAUGE	ENAMEL	DOUBLE COVERED COTTON D.C.C.	SINGLE COVERED COTTON S.C.C.	DIAMETER (INCHES)	CURRENT CARRYING CAPACITY @700 CM PER AMP	OHMS PER 1,000 FT. @20° C
0000	—	—	—	.4600	302.3	0.04901
000	—	—	—	.4096	239.7	0.06180
00	—	—	—	.3648	190.1	0.7793
0	—	—	—	.3249	150.7	0.09827
1	—	3.3	3.3	.2893	119.6	0.1239
2	—	3.6	3.8	.2576	94.8	0.1563
3	—	4.0	4.2	.2294	75.2	0.1970
4	—	4.5	4.7	.2043	59.6	0.2485
5	—	5.0	5.2	.1819	47.3	0.3133
6	—	5.6	5.9	.1620	37.5	0.3951
7	—	6.2	6.5	.1443	29.7	0.4982
8	7.6	7.1	7.4	.1285	23.6	0.6282
9	8.6	7.8	8.2	.1144	18.7	0.7921
10	9.6	8.9	9.3	.1019	14.8	0.9989
11	10.7	9.8	10.3	.09074	11.8	1.260
12	12.0	10.9	11.5	.08081	9.33	1.588
13	13.5	12.0	12.8	0.7196	7.40	2.003
14	15.0	13.8	14.2	.06408	5.87	2.525
15	16.8	14.7	15.8	.05707	4.65	3.184
16	18.9	16.4	17.9	.05082	3.69	4.016
17	21.2	18.1	19.9	0.4526	2.93	5.064
18	23.6	19.8	22.0	0.4030	2.32	6.385
19	26.4	21.8	24.4	.03589	1.84	8.051
20	29.4	23.8	27.0	.03196	1.46	10.15
21	33.1	26.0	29.8	.02846	1.16	12.80
22	37.0	30.0	34.1	0.2535	.918	16.14
23	41.3	31.6	37.6	.02257	.728	20.36
24	46.3	35.6	41.5	.02010	.577	25.67
25	51.7	38.6	45.6	.01790	.458	32.37
26	58.0	41.8	50.2	.01594	.363	40.81
27	64.9	45.0	55.0	.01420	.288	51.47
28	72.7	48.5	60.2	.01264	.228	64.90
29	81.6	51.8	65.4	.01126	.181	81.83
30	90.5	55.5	71.5	.01003	144	103.2
31	101.0	59.2	77.5	.008928	.114	130.1

32	113.0	62.6	83.6	.007950	.090	164.1
33	127.0	66.3	90.3	.007080	.072	206.9
34	143.0	70.0	97.0	.006305	.057	260.9
35	158.0	73.5	104.0	.005615	.045	329.0
36	175.0	77.0	111.0	.005000	.036	414.8
37	198.0	80.3	118.0	.004453	.028	523.1
38	224.0	83.6	126.0	.003965	.022	659.6
39	248.0	86.6	133.0	.003531	.018	831.8
40	282.0	89.7	140.0	.003145	.014	1,049.0

Appendix D

ELECTRONIC AND ELECTRICAL ABBREVIATIONS

TERM	ABBREVIATION
ADJUSTABLE	adj
ALTERNATING CURRENT	ac
AMBIENT	amb
AMPERE	A
AMPERE-HOUR	Ah
AMPLITUDE MODULATION	AM
ANTILOGARITHM	antilog
APPROXIMATE, –LY	Approx
ATMOSPHERE	atm
ATTO – (10^{-18})	a
AUDIO FREQUENCY	af
AUTOMATIC FREQUENCY CONTROL	afc
AUTOMATIC GAIN CONTROL	agc
AUTOMATIC VOLUME CONTROL	avc
AVERAGE	avg
BEAT-FREQUENCY OSCILLATOR	bfo
BEL	B
BINARY CODED DECIMAL	BCD
BITS PER SECOND	b/s
BRITISH THERMAL UNIT	Btu
BROADCAST	BC
CALIBRATE, CALIBRATION	cal
CALORIE	cal
CATHODE-RAY OSCILLOSCOPE	CRO
CATHODE-RAY TUBE	CRT
CENTI (10^{-2})	c
CENTIGRAM	cg
CENTIMETER	cm
CENTIMETER-GRAM-SECOND	cgs
CIRCULAR MIL	cmil
CLOCKWISE, CONTINUOUS WAVE	cw
COMPLEMENTARY METAL-OXIDE SEMICONDUCTOR	CMOS
COSECANT	cs
COSINE	cos
COTANGENT	cot
COULOMB	C
COUNTERCLOCKWISE	ccw
CUBIC CENTIMETER	cm^3
CUBIC FOOT	ft^3
CUBIC FOOT PER MINUTE	ft^3/min
CUBIC FOOT PER SECOND	ft^3/s
CUBIC INCH	$in.^3$
CUBIC METER	m^3
CUBIC METER PER SECOND	m^3/s
CYCLE PER SECOND	c/s
DEKA- (10)	da
DECI- (10^{-1})	d
DECIBEL	dB
DECIBEL REFERRED TO 1 MILLIWATT	dBm
DECIBEL REFERRED TO 1 WATT	dBw
DECIBEL REFERRED TO 1 VOLT	dBv
DEGREE CELSIUS	°C
DEGREE FAHRENHEIT	°F
DEGREE KELVIN	°K
DIAMETER	dia
DIGITAL VOLTMETER	dvm
DIODE-TRANSISTOR LOGIC	DTL
DIRECT CURRENT	dc
DIRECT-CURRENT WORKING VOLTS	dcwv
DOUBLE-POLE, DOUBLE-THROW	dpdt
DOUBLE-POLE, SINGLE-THROW	dpst
ELECTROMOTIVE FORCE	emf
ELECTRONVOLT	eV
EQUATION	Eq.
EXTERNAL	ext
FARAD	F
FIELD-EFFECT TRANSISTOR	FET
FILAMENT	fil
FOOT	ft
FOOT PER SECOND	ft/s
FOOT POUNDAL	ft-pdl
FOOT-POUND FORCE	ft-lbf
FOOT-SECOND	ft-s
FREQUENCY MODULATION	FM
GAUSS	G
GIGA – (10^9)	G
GIGACYCLES PER SECOND	Gc/s
GIGAELECTRON VOLT	GeV
GIGAHERTZ	GHz
GILBERT	Σ
GRAM	g
GRAVITY	g
GROUND	gnd
HECTO-(10^2)	h
HENRY	H
HERTZ	Hz
HIGH FREQUENCY	hf
HORSEPOWER	hp
HOUR	h
INCH	in.
INCH PER SECOND	in./s

TERM	ABBREVIATION
INFRARED	IR
INSIDE DIAMETER	ID
INSULATED-GATE FIELD-EFFECT TRANSISTOR	IGFET
INTEGRATED CIRCUIT	IC
INTERMEDIATE FREQUENCY	i-f
KILO-(10^3)	k
KILOCYCLE PER SECOND	kc/s
KILOGRAM	kg
KILOHERTZ	kHz
KILOHM	kΩ
KILOMETER	km
KILOVAR	kvar
KILOVOLT	kV
KILOVOLTAMPERE	kVA
KILOWATT	kW
KILOWAT-HOUR	kWh
LAMBERT	Lb
LARGE-SCALE INTEGRATION	LSI
LOGARITHM	log
LOW FREQUENCY	lf
MAXIMUM	max
MAXWELL	Mx
MEDIUM-SCALE INTEGRATION	MSI
MEGA- (10^6)	M
MEGACYCLE PER SECOND	Mc/s
MEGAHERTZ	MHz
MEGAVOLT	MV
MEGAWATT	MW
MEGOHM	MΩ
METAL-OXIDE SEMICONDUCTOR	MOS
METAL-OXIDE SEMICONDUCTOR FIELD-EFFECT TRANSISTOR	MOSFET
METER	m
MICRO- (10^{-6})	μ
MICROAMPERE	μA
MICROFARAD	μF
MICROHENRY	μH
MICROHM	$\mu\Omega$
MICROMETEr, micron	μm
MICROSECOND	μs
MICROVOLT	μV
MICROWATT	μW
MILLI- (10^{-3})	m
MILLIAMPERE	mA
MILLIBAR	mnb
MILLIGRAM	mg
MILLIHENRY	mH
MILLIMETER	mm
MILLISIEMENS	mS
MILLIMICRON	(see nanometer)
MILLIOHM	mΩ
MILLISECOND	ms
MILLIVOLT	mV

TERM	ABBREVIATION
MILLIWATT	mW
MINIMUM, MINUTE	min
NANO - (10^{-9})	n
NANOAMPERE	nA
NANOFARAD	nF
NANOHENRY	nH
NANOMETER	nm
NANOSECOND	ns
NANOWATT	nW
NEGATIVE	neg
NEGATIVE-POSITIVE-NEGATIVE	NPN
NEWTON	N
NORMALLY-CLOSED	NC
NORMALLY-OPEN	NO
NUMBER	No.
OERSTED	Oe
OHM	Ω
OUNCE	oz
OUTSIDE DIAMETER	OD
PARTS PER MILLION	ppm
PEAK	pk
PEAK INVERSE VOLTAGE	PIV
PEAK-TO-PEAK	P-P
PICO-(10^{-12})	p
PICOAMPERE	pA
PICOFARAD	pF
PICOSECOND	ps
PICOWATT	pW
POSITIVE	pos
POSITIVE-NEGATIVE-POSITIVE	PNP
POTENTIOMETER	pot
POUND	lb
POUND PER SQUARE FOOT	lb/ft^2
POUND PER SQUARE INCH	$lb/in.^2$
POWER FACTOR	pf
PULSE-AMPLITUDE MODULATION	PAM
PULSE-CODE MODULATION	PCM
PULSE-DURATION MODULATION	PDM
PULSE-POSITION MODULATION	PPM
PULSE REPETITION FREQUENCY	prf
PULSE-WIDTH MODULATION	PWM
RADIAN	rad
RADIO FREQUENCY	rf
REACTIVE VOLTAMPERE	(see var)
RECEIVER	rcvr
REFERENCE	ref
ROOT-MEAN-SQUARE	rms
SECANT	sec
SECOND	s
SENSITIVITY	sens
SIEMENS	S
SILICON-CONTROLLED RECTIFIER	SCR
SINE	sin
SINGLE-POLE DOUBLE-THROW	spdt

TERM	ABBREVIATION
SINGLE-POLE SINGLE-THROW	spst
SINGLE SIDEBAND	SSB
SMALL-SCALE INTEGRATION	SSI
SQUARE	sq
SQUARE FOOT	$ft.^2$
SQUARE INCH	$in.^2$
SQUARE METER	m^2
STANDARD	std
STANDING-WAVE RATIO	SWR
STERADIAN	sr
SYNCHRONOUS, SYNCHRONIZING	sync
TANGENT	tan
TELEVISION	TV
TEMPERATURE	temp
TERA-(10^{12})	T
TERACYCLE PER SECOND	Tc/s
TERAHERTZ	THz
TESLA	T
TRANSISTOR-TRANSISTOR LOGIC	TTL
TRANSISTOR-VOLTMETER	tvm
TRANSMIT-RECEIVE	T-R
TRAVELING-WAVE TUBE	TWT
ULTRA-HIGH FREQUENCY	uhf
ULTRAVIOLET	UV
VACUUM-TUBE VOLTMETER	vtvm
VAR (REACTIVE VOLTAMPERE)	VAR
VARIABLE-FREQUENCY OSCILLATOR	VFO
VERSUS	vs
VERY HIGH FREQUENCY	vhf
VERY-LARGE-SCALE INTEGRATION	VLSI
VERY LOW FREQUENCY	vlf
VOLT	V
VOLTAGE STANDING-WAVE RATIO	vswr
VOLTAMPERE	VA
WATT	W
WATTHOUR	WH
WEBER	Wb

Glossary

ac—Abbreviation for alternating current.

agc—Automatic gain control. The circuit samples demodulated gain levels and provides an automatic correction bias which maintains a predetermined signal amplitude.

air core—The term used to describe inductors having no magnetic core material.

align—To tune or adjust a circuit to meet specific requirements.

alternating current—Current which has periodic alternations of positive and negative polarities.

alternator—A device which produces alternating current.

ammeter—An instrument that measures the rate of current flow.

ampere—A unit of current.

amplification—The process of increasing the current, voltage, or power of a signal.

amplification factor—An indication of the general amplification characteristics of a vacuum tube, defined as the ratio of the change in plate voltage to a small change in the grid voltage, with the plate current remaining at a constant level. The symbol for amplification factor is the Greek letter μ.

amplifier—A device designed to increase the signal voltage, current, or other waveform measured in either a positive or negative direction.

amplitude modulation (AM)—The type of modulation commonly used for "standard" radio broadcasting. The carrier signal is modulated by low-frequency audio signals so that the overall waveform amplitude varies above and below the normal carrier level at a rate and amplitude change corresponding to the modulating signal.

analog electronics—That part of electronics that deals with uniformly changing signals such as a sine wave. See *digital electronics*.

anode—The plate or positive electrode.

antenna—A device used for receiving or transmitting signals. Sometimes called an aerial.

aquadag—The graphite coating in a cathode-ray or television tube.

attenuate—To diminish the amplitude of a signal.

attenuator—A device for reducing signal amplitude by using either fixed or variable components.

audio amplification—The increase of signal amplitude within the audible frequency range.

autotransformer—A single-coil transformer where primary and secondary are connected together in one winding.

avc—Automatic volume control.
average value—The average of all instantaneous values of I or E in one half cycle. Can be denoted as the product of 0.636 times the peak amplitude.

bandpass filter—An electronic network which passes a specific band of frequencies.
base terminal—An electrode of a transistor. Usually compared to the grid of a vacuum tube.
beta—The current gain in a grounded-emitter transistor amplifier. The symbol for a current gain is the Greek letter B.
bfo—Beat-frequency oscillator.
bias—The difference of potential applied between the grid and cathode of a tube or between transistor elements to provide an operating point at zero signal input.
blanking—Electron-beam cut-off in a cathode-ray tube during beam retrace time.
bleeder—One or more resistors shunting the output of a power supply to improve voltage regulation by providing a fixed current drain.
broadcast band—The term generally applied to the rf frequency span between 550 kHz and 1600 kHz allocated to "standard" radio transmission.
buffer stage—An amplifier or other circuit usually employed in rf amplifier (transmitter) stages to isolate the oscillator and subsequent amplifiers.
bus bar—A primary power distribution point connected to the main power source.

capacitance—The quantity of electric charge (usually in fractional farad quantities) which a capacitor is capable of storing a given voltage.
capacitive reactance—The opposition which a capacitor offers to ac at a specific signal frequency.
capacitor—A device capable of "storing" electrons between two conducting surfaces insulated by a dielectric.
carrier—An rf signal capable of being modulated to carry information.
cascade—Circuits or stages connected in sequence.
cathode—The element in a vacuum tube which emits electrons.
cathode follower—A tube circuit where the output signal appears across the cathode with the anode at signal ground. Also a grounded-collector transistor circuit.
cathode-ray tube—A tube with a phosphor-treated face on which an electron beam traces an image.
center frequency—The term usually applied to the unmodulated FM carrier frequency.
characteristic impedance—The impedance of a transmission line or attenuator network.
charge—The quantity of energy stored by a capacitor or storage-type battery.
choke coil—An inductor designed to provide a high impedance to ac.
chopper—A circuit which converts dc to ac by periodic interruption (chopping) of the dc.
circuit breaker—An electromagnetic or thermal device that opens a circuit when the current exceeds a certain value.
clamper—A circuit designed to restore the dc component of a signal waveform.
class A amplifier—An amplifier biased to operate on the linear portion of the characteristic curve.
class B amplifier—An amplifier biased to operate at or near the tube or transistor cut-off point. Positive alternations of the input signal cause current flow.
class C amplifier—An amplifier biased beyond the cut-off point so current flows for only a portion of the positive alternations of the input signal.
clipper—A circuit designed to remove portions of the input signal amplitude.
coaxial cable—A transmission line consisting of two concentric conductors insulated from each other.

cold-cathode tube—A tube which requires no external heat current source to produce electron emission.

color codes—The identification of electronic components by color bands or dots which relate to numerical values. (See Appendix B).

colpitts oscillator—An oscillator using series capacitors across the resonant circuit inductor.

complementary metal-oxide semiconductor—A semiconductor device that consists of two complementary *MOSFETs* in a single package. Abbreviated CMOS.

condenser—See Capacitor.

conductance—The current-carrying ability of a wire. The unit value is siemens and is the reciprocal of resistance.

conductor—A medium which carries a flow of electric current.

continuous wave—An unmodulated rf waveform of constant amplitude. Usually the term applied to a wave transmitted in bursts of short and long duration to form the Morse code.

control grid—The grid to which a signal is usually applied in a tube.

conventional current flow—The theory that current flows from positive to negative.

converter—The stage in a superheterodyne receiver which produces the i-f signal by mixing the rf carrier with a locally-generated signal.

counter emf—Counter electromotive force; an emf induced in a coil or armature in opposition to the applied voltage.

counting circuit—A circuit which produces a voltage in proportion to the frequency of uniform input pulses.

coupling—The effective linkage connecting two electronic circuits; usually transformers, capacitors, and inductors.

CRO—Cathode-ray oscilloscope.

cross-modulation—Modulation of a desired signal by an unwanted signal.

crossover frequency—The frequency in a multiple-speaker where the signal is divided and fed to high- and low-frequency systems.

crystal oscillator—A signal-generating circuit in which the frequency is controlled by a piezo-quartz crystal.

current limiter—A fuse-like protective device designed to limit current flow in a circuit.

cut-off frequency—The frequency of a filter or other circuit beyond which signal flow ceases.

CW—See *continuous waves*.

cycle—In ac, one complete alternation, positive and negative.

dc amplifier—An amplifier using direct coupling (no coupling capacitors or transformers).

dc restorer—A clamper circuit which restores the dc level to a signal waveform.

decibel—One tenth of a bel.

decoupling circuit—A resistance-capacitance circuit which isolates signal-carrying circuits from circuits common to other signal-carrying circuits.

de-emphasis circuit—An RC filter used after FM detector systems to decrease high-frequency signal levels which were increased during transmission.

delayed avc—An automatic volume-control circuit designed to produce an avc bias only for signals above a fixed amplitude.

demodulation—A signal-rectifying system which extracts the modulating-signal component from the modulated carrier.

detection—To separate modulation from the signal.

deviation ratio—The ratio of maximum FM carrier deviation to the highest frequency audio-modulating signal employed.

dielectric—The insulating material between the two conductors, such as in a capacitor, or the insulating material between transmission line conductors.
digital electronics—That part of electronics that deals with signals having discrete values. See *analog electronics*.
diode—A two-element tube or two-terminal solid-state rectifier.
direct current—Current flow in one direction.
discriminator—The detector used in frequency modulation. It is used also to compare two ac signals.
distortion—Unwanted modification of a desired signal.
doubler—A circuit in transmitting systems which doubles the frequency of the input signal. In power supply systems, a circuit for doubling voltage amplitude.
driver stage—An audio or rf amplifier stage preceding the final or power amplifier.

eddy currents —Stray, induced currents in a conducting material caused by a varying magnetic field.
effective value—The value of alternating current or voltage equal to the product of 0.707 times the peak amplitude. Root mean square value.
efficiency—The ratio of output power to input power, generally expressed as a percentage.
electrode—A terminal used to emit, collect, or control electrons.
electrolyte—A solution or a substance which is capable of conducting electricity; it may be in the form of either a liquid or a paste.
electrolytic capacitor—A capacitor utilizing an electrolyte to form the dielectric insulation.
electromagnet—A magnet made by passing current through a coil of wire wound on a soft iron core.
electromotive force (emf)—The force that produces a current in a circuit (voltage).
electronic switch—A circuit which introduces a start-stop action by electronic means.
emitter—Transistor electrode similar, functionally, to the cathode of a tube.

farad—The unit of capacitance. Fractional values are used in practical electronics.
feedback—A transfer of energy from the output of a circuit or component back to its input.
feedback oscillator—A signal-generating circuit which employs regenerative feedback to sustain oscillations.
ferrite—A metallic compound used for high-Q core materials in inductors.
field-effect transistor (FET)—A semiconductor amplifying device in which the gate controls the flow of current through the channel to the source and drain connections.
filament—The electrode in a vacuum tube which is heated for electron emission or which transfers its heat to a separate cathode.
filter—A circuit designed to pass certain signal components and attenuate others.
filter capacitor—An electrolytic capacitor used in power supplies to reduce ripple.
filter choke—An inductor used in power supplies to reduce ripple.
flip-flop—A bistable circuit which can be triggered to its other state by an input signal or pulse.
forward bias—The bias applied between the base and emitter of a transistor or through a diode.
frequency—The number of complete cycles per second in an alternating ave. A cycle includes negative and positive "excursions." Denoted in hertz (Hz).
frequency division—A circuit designed to reduce the repetition rate of pulse waveforms or decrease the frequency of ac signals.

frequency modulation—A system where the frequency of the carrier signal is shifted above and below its normal center frequency by the modulating signal.

full-wave rectifier—A power supply circuit which uses both alternations of the ac waveform to produce direct current.

gain—The ratio of the output signal to the input signal, voltage or current. Denoted as A.

gallium arsenide—A semiconductor material composed of a compound of gallium and arsenic.

galvanometer—An instrument used to measure small dc currents.

gas tube—A tube containing gas which must ionize before conduction can occur.

generator—A machine that changes mechanical energy into electrical energy by rotating coils of wire within a fixed magnetic field.

grid—A wire, usually in the form of a spiral, used to control the electron flow in a vacuum tube.

grounded base—A transistor amplifier circuit similar to a grounded-grid tube circuit.

grounded collector—A transistor circuit similar to a cathode-follower tube circuit.

grounded emitter—A transistor circuit similar to the conventional grounded-cathode tube amplifier.

grounded grid—A tube circuit with the control grid at signal ground.

half-wave rectifier—A tube or solid-state diode which converts ac to pulsating dc by rectifying one alternation of each ac cycle.

harmonic—A signal related to a fundamental signal by some multiple.

heater—A vacuum-tube electrode which heats the cathode.

henry—The basic unit of inductance. One henry represents the amount of inductance present when a current change of 1 amp per sec produces an induced voltage of 1 V.

heptode—A tube with seven electrodes.

heterodyne—The electronic mixing of two signals of different frequencies to produce a third signal.

hexode—A tube with six electrodes.

high fidelity—An audio system which reproduces the full audio-frequency spectrum with negligible distortion.

high-pass filter—A circuit that transfers high-frequency signals while attenuating the lows.

hole—In semiconductors, the space left vacant in an atom by a departed electron. Holes "flow" in a direction opposite to that of electrons and bear a positive charge.

hysteresis—The phenomenon present in magnetic materials where the flux density (B) lags the magnetizing force (H).

impedance—A combination of resistance and reactance which opposes ac current flow.

inductance—The property of a coil which opposes a change in current.

induction—The process of inducing a potential or magnetization in another component by magnetic lines of force.

inductive reactance—The opposition an inductor offers to ac for a given signal frequency. It is measured in ohms.

inductor—A circuit element designed so that inductance is its most important property, such as a coil.

in phase—The condition that exists when two ac waves of the same frequency pass through their maximum and minimum values of like polarity at the same instant in time.

integrated circuit—An electronic device that consists of many components on a thin chip of semiconductor.

intermediate frequency—The signal obtained by heterodyning or mixing two signals of different frequencies. Denoted as i-f.

kilo—A prefix meaning 1,000.

kilocycles (kc)—One thousand cycles. (The preferred unit now is kilohertz.)

Kirchhoff's current law—The basic law which states that the sum of currents flowing into any junction of an electric circuit is equal to the sum of currents flowing out of that junction.

Kirchhoff's voltage law—The basic law which states that the sum of voltage sources around any closed circuit is equal to the sum of the individual voltage drops across the resistances of the circuit.

lag—The amount (in degrees) one ac wave is behind another in time.

laminated core—A core built up from thin sheets of metal; used in transformers and relays.

lead—The opposite of lag. Also, a connecting wire.

level control—A variable control for adjusting signal levels.

limiter—A circuit which limits the peak amplitudes of signal waveforms to a predetermined level.

linear—A circuit where the output signal varies in direct proportion to the input.

load—A resistor or transformer, usually, across which the output signal of a circuit is developed.

low-pass filter—A circuit designed to pass low-frequency signals and attenuate the highs.

magnetic field—The area in which magnetic lines of force exist.

megohm—A million ohms.

metal-oxide silicon field-effect transistor—A special type of field-effect transistor that uses a thin metal film that is insulted from the semiconductor material by a thin oxide film. Known as MOSFET.

micro—A prefix meaning one-millionth.

milliammeter—An ammeter constructed to measure fractional (thousandths) values of an ampere.

modulation—The process of modifying an rf carrier signal to transmit audio or video signal information over great distances.

mutual inductance—The inductance (coupling) established when two coils are close together.

null—Zero or minimum.

octode—A tube with eight electrodes.

ohm—The unit of resistance.

ohmmeter—An instrument which measures resistance in ohms.

Ohm's law—A basic law of electricity establishing the mathematical relationships between current, voltage, and resistance values (E equals I times R).

oscillator—A regenerative circuit designed to produce signals.

oscilloscope—An instrument using a cathode-ray tube which presents a visual display of electric signals or waveforms.

overmodulation—The modulation of an rf carrier in excess of 100%.

pad—An attenuator circuit usually used as a coupling between circuits where the output of one is too high for the other.
peak inverse voltage—The peak voltage a rectifier will handle without arcing internally with a polarity opposite to that causing conduction.
peak-to-peak value—The over-all amplitude of a signal measured from it lowest (or most negative) peak to its highest (or most positive) peak.
peak value—The instantaneous maximum value of a waveform or signal.
pentagrid converter—A five-element tube used as a mixeroscillator in superheterodyne receivers.
phase angle—The angle of lead or lag between voltage and current in an ac waveform.
plate resistance—The dynamic (ac signal) resistance of a tube.
potentiometer—A variable resistor.
power amplifier—An audio or rf amplifier designed to deliver signal energy (power) rather than signal voltage.
power factor—The cosine of the phase angle between voltage and current; an efficiency rating.
power supply—A circuit designed to furnish operating voltages and currents for electronic devices.
preamplifier—An additional stage of amplification preceding another amplifier to increase signal amplitudes above a given level.
push-pull circuit—An amplifier circuit with two tubes or transistors operating so that when one is conducting on a positive alternation, the other operates on a negative alternation.

Q—A symbol for the figure merit of an inductor.
quiescent operating point—Zero-signal operation of a device.

radio-frequency amplifier—An amplifier designed to increase rf signal levels.
radio-frequency choke—An inductor designed to introduce reactance when used in series with a signal-carrying lead.
ratio—The value obtained by dividing one number by another, indicating their relative proportional relationship.
ratio detector—A dual-diode frequency-modulation detector.
reactance—The opposition to ac current flow by an inductor or capacitor.
rectifier—A device that changes alternating current to direct current.
regulation—The degree by which voltage is held near its no-load value when a load is applied.
relay—An electromechanical device used to remotely open or close a circuit.
reluctance—The opposition offered by a material to magnetic lines of force.
resistance—The opposition to current flow; measured in ohms.
resonance—A condition in a tuned circuit where reactances cancel at a specific frequency.
reverberation—Sound waves reinforced by reflection.

saturation point—In a tube or transistor an increase in voltage produces no further (or very little) increase in current.
sawtooth voltage—A waveform characterized by a gradual rise and rapid decline of amplitude.
screen grid—The electron-beam accelerating electrode in a tube.
secondary emission—Electron emission from an electrode other than the cathode in a tube when struck by a high-velocity electron beam.

self-bias—A circuit that produces bias within its associated circuits as a result of internal current flow.
semiconductor—Resistivity value lies between that of a conductor or insulator, such as a transistor or diode.
siemens—The unit of conductance, which is the reciprocal of the ohm. This term replaces the now obsolete term mho (ohm spelled backwards).
silicon-controlled rectifier—A solid-state rectifier in which conduction can be started by applying a control voltage.
solenoid—An electromagnetic coil with a movable plunger.
standing wave—Current and voltage waves on a transmission line formed by a reflection of the desired signal, caused by impedance mismatch.
static—A fixed condition; no motion.
stereophonic—Sound reproduction utilizing two or more separate amplification channels feeding respective loudspeaker systems.
superheterodyne receiver—A radio or TV receiver employing a mixer stage to produce an intermediate frequency on all incoming signals; the i-f is amplified by fixed-selectivity circuits.
suppressor grid—A grid (usually between anode and screen grid) at ground potential to eliminate secondary emission.
sweep generator—A signal generator whose output signal is varied (swept) through a given frequency range.

tachometer—An instrument which indicates revolutions per minute.
tank circuit—Usually refers to a parallel resonant circuit.
tetrode—A four-electrode tube or transistor.
thermionic emission—Production of electron emission by heat.
thermistor—A resistor that change its resistance value to compensate for temperature changes.
thermocouple—A junction of two dissimilar metals that produces a voltage when heated.
time constant—The product of R and C in a series circuit.
trace—A visible line on the screen of a cathode-ray tube.
transducer—A device for converting energy from one form to another, such as vibrations from a phonograph pickup into audible sounds.
transformer—A device with two or more coils linked by magnetic lines of force used to transfer energy from one circuit to another and can increase or decrease (step up or step down) the voltage.
transmission lines—Conductors used to carry energy from a source to a load.
transient—An irregular signal of fractional duration as compared to the primary signal.
transistor—A solid-state device.
trigger—A pulse employed to start or stop the operation of a circuit or device.
triode—A three-electrode vacuum tube with a cathode, control grid, and plate.
tuned circuit—A circuit at resonance.
turns ratio—The ratio of primary-to-secondary turns of a transformer.

unmodulated—An rf carrier signal with no modulation.

vacuum tube—An evacuated envelope containing two or more electrodes.
vacuum-tube voltmeter—A high input impedance test instrument with a tube (or transistor) circuit.
varactor—A semiconductor variable capacitor controlled by a dc voltage.

variable-frequency oscillator—A signal-generating circuit in which component values can be varied to alter the frequency of the output signal.
vector—A line used to represent both direction and magnitude of ac signals.
video amplifier—A circuit capable of amplifying a very wide range of frequencies, from the audio band and higher.
volt—The unit of electrical potential (emf).
voltage divider—Resistors placed in series across a voltage to obtain intermediate values of voltage.
voltage doubler—A power-supply circuit so designed that the rectified voltage amplitude is almost double the input ac amplitude.

watt—The unit of electric energy or power.
waveform—The shape of the wave obtained when instantaneous values of an ac quantity are plotted against time in rectangular coordinates.
wavelength—The distance, usually expressed in meters, traveled by a wave during the time interval of one complete cycle. It is equal to the velocity divided by the frequency.
working voltage—The maximum voltage at which a device will operate continuously with safety.

yagi antenna—An antenna system employing a basic antenna element as well as reflector and director rods.
yoke—In a television receiver, a coil arrangement around the neck of the picture tube which provides electromagnetic deflection of the CRT beam vertically and horizontally.

zener diode—A solid-state semiconductor that has voltage-regulation characteristics when subjected to reverse bias.
zero bias—The absence of a potential between the grid and cathode of a tube, or between the emitter-base or other electrodes of a transistor.

Index

A

abbreviations, 200-202
ac power supply (TV), 141
alkaline battery, 16
AM radio
  basic theory of, 91
  receiver block diagram for, 94
  technical description of, 95
amplifiers, 98, 102
amplitude modulation (AM), 92
AND gates, 177, 179
anodes, 57
APC, 138
armature, 20
audio amplifier, 98
audio cable, 74
audio connectors, 71
audio processing (TV), 121
audio signals, 91, 92
automatic CRT tracking, 128
automatic gain control (AGC), 98, 118

B

band-pass filter, 27
barrier region, diode, 48
basic components, 1-24
bathtub capacitor, 8
batteries, 15-16
beam power tubes, 61
bimetal strip circuit breaker, 24
binary number systems, 175
block diagrams, 79
  drawing of, 84
  schematic diagram and, 86
Boolean algebra, 177
breadboards, 40-41
buffer gate, 182

C

cable
  audio, 74
  coaxial, 75
  speaker wire, 76
  special-purpose, 73
  twin-lead, 75
calibrator, 173
capacitance, 6
capacitors, 6-10
  color coding for, 195
case configuration, transistor, 45
cathode-ray tubes (CRT), 64
  color, 65
  electrostatic vs. magnetic deflection in, 64
cathodes, 57
character generator (television), 107
chokes, 10
chroma circuits, 121
circuit breakers, 23-24
circuit components, 1
circuit diagrams, drawing of, 85
coaxial cable, 75
  connectors for, 72
coaxial speakers, 36
coils, 10
  schematic symbols for, 11, 87
color coding, 195-197
components
  numbering of, 88
  standard abbreviations for, 89
connectors
  audio, 71
  coaxial cable, 72
  multiple-wire, 69
  rf, 72
control circuits, 164, 168
control grid, 58
crosshatch generator, 147
CRT blanking, 172
CRT grid voltage generation, 130

CRT tracking systems, 101
crystals, 25

D

decimal number systems, 174
degaussing circuit, 101
demodulation, 94
detectors, 95
diac, 52
diagrams, 77-90
digital circuits, 174-185
  measuring logic levels in, 185
  reading schematics for, 183
digital gates, 178
digital integrated circuits, 178
diodes, 46
  light-emitting (LEDs), 54
  tunnel, 48
  unilateral, 46
  vacuum tube, 58
  varactor, 48
  zener, 47
directional microphones, 32
DPDT switch, 18
drivers, 98
  in-phase, 139
dry-cell batteries, 16
dual in-line integrated circuits, 50

E

electrodes, vacuum tube, 57
electrolytic capacitors, 10
electronic schematic symbols, 186-194
exclusive-OR gate, 182
exploded views diagram, 83

F

feedthrough capacitor, 8
field-effect transistors (FETs), 52
filters, 27
  mechanical, 28
  networks of, 27, 28
  video, 102
fixed capacitor, 6, 8
fixed resistor, 3
frequency modulation (FM), 92
frequency spectrum, 93
full-wave rectifier, 58
fuses, 22-23

G

ganged capacitor, 7
ganged variable, 8
gates, digital, 178

H

half-wave rectifier, 58
headphones, 29
heads, tape recorder, 37
Heathkit 19-inch color TV with remote (GR-1903), 100-150
Heathkit dual-trace oscilloscope (IO-4210), 151-173
Heathkit portable AM radio (GR-1009), 91-99
heatsinks, 38, 39, 40
high-pass filter, 27
horizontal circuit, 170
horizontal sweep system, 138
horizontal systems (TV), 137

I

i-f amplifiers, 95, 114
i-f module, stereo, 118
in-phase driver, 139
indicating components, 25
inductance, 10
inductor, 10
infrared emitter and detector, 56
insulated-gate field-effect transistor (IGFET), 52, 53
integrated circuits, 49
  block diagram for, 49
  digital, 178
  dual in-line, 50
inter-unit cabling, 70
interconnecting devices, 66-76, 66
international system of units (SI), 87
inverters, 181
ionization, 30
IR transmitter (TV), 146

J

jacks, 99
junction field-effect transistor (JFET), 52
junctions, diode, 46

L

lamps, 30
layout diagrams, 80
light-emitting diodes (LEDs), 54
liquid-crystal displays (LCDs), 55
logic levels, 185
logic probe, 185
logic schematic, 183
low-pass filter, 27
luminance circuits, 127

M

magnetic circuit breakers, 24
master scan oscillator, 138
mechanical filters, 28
mechanical construction diagrams, 82
metal-oxide semiconductor (MOSFET), 52, 53
meters, 31
microphones, 32-34
microprocessor, 112
mid-range speakers, 36
millifarad, 7
miscellaneous components, 25
mixer, 95
modulation of audio signals, 92
modulator, 92
multiconductor ribbon cable, 73
multiple-wire connectors, 69
multiplier resistors, 31
multivibrator start-up system (TV), 130

N

n-type semiconductors, 42
NAND gates, 181
nematic liquids, 55
neon lamps, 30
nickel-cadmium (rechargable) battery, 16, 17
NOR gates, 181
normally closed and normally open relay, 21
NOT function, 178, 183
number systems, 174-177

O

ohms, 3
OR function, 177, 180
oscillators, 92, 95
oscilloscope, block diagram of, 153

P

p-type semiconductors, 42
parts lists, 78
pentagrid tubes, 62
pentode, 60
phase-locked loop tuning, 101
phono pickups, 34
photocells, 55
photovoltaic cells, 55
pictorial diagrams, 81
piezoelectricity, 25
polarized capacitors, 10
potentiometer, 5
power amp, 140

power supply, 171
block diagram of, 84
circuit diagram of, 86
schematic diagram of, 78
primary transformer, 13
printed-circuit boards, 66, 67
product detector, 62
pulse width modulator, 139
push-pull output amplifier, 98

R

radio
basic theory of AM, 91
receivers for, 94
stations for, 92
radio and TV schematics, 91-150
radio receiver, 84
ramp generator, 136
rectifiers, half- and full-wave, 58
regulator tubes, 63
relay, 20
NO and NC, 21
schematic symbols for, 20
remote control, 101, 141
resistance, 3
resistors, 1-5
color coding for, 2, 196
rf connectors, 72
rheostat, 5
rotary switch, 19

S

schematic diagrams, 77
standard units of measurement in, 89
schematic symbols, proper use of, 87
screen grid, 60
second audio (TV), 121
secondary emission, 60
secondary transformer, 13
semiconductors, 42
metal-oxide (MOSFET), 52, 53
n- and p-type, 42
shielding, 69
shunted meters, 31
shunting, 69
silicon-controlled rectifier (SCR), 51
silicon-controlled switch (SCS), 51
slide switch, 19
slow blow fuse, 23
solar cells, 55
solid-state devices, 42-56
sound waves, 91
SPDT switch, 18
speaker wire, 76
speakers, 34-36
specialized electronic equipment, 151-173
standard noisy signal, 136
stereo, 102
stereo decoder module, 120
stereo i-f module, 118
substrate, 53
superheterodyne receiver, 94
suppressor, 60
sweep circuits, 164, 166
switches, 18-19
sync separator, filter, and differentiator, 138

T

tape recorder heads, 37
television, 100
schematics for, 91-150
twin-lead cable use in, 75
use of cathode-ray tubes in, 64
terms and abbreviations, 200-202
tetrode, 60
thermistors, 56
toggle switch, 19
traces, 66
transducers, 25
microphone use of, 34
transformers, 12
applications for, 14
color coding for, 197
primary and secondary, 13
schematic symbols for, 13
transistors, 42-46
circuits using, 43
field-effect (FET), 52
insulated-gate field effect (IGFET), 52, 53
junction field-effect (JFET), 52
package type or case configuration of, 45
schematic symbols for, 43
unijunction, 54
triac, 52
triaxial speakers, 36
trigger amplifier, 163
trigger circuit, 164, 165
trim pots, 5
trimmer capacitor, 8
trimmer resistors, 4
triode vacuum tube, 59
truth table, 180
tuner module, 102
tuner-rf amplifier, 95
tuning systems (TV), 106
tunnel diodes, 48
tweeter speakers, 36
twin-lead cable, 75

U

uhf television signals, 106
unidirectional microphones, 32
unijunction transistor (UJT), 54
unilateral diodes, 46

V

vacuum tubes, 57-65
varactor diodes, 48
variable capacitor, 6
vertical circuits, 153
vertical countdown system, 135
vertical deflection, 163
vertical integrator, 135
vertical output circuits (TV), 131
vertical ramp generator and comparator, 136
vhf television signals, 106
video filters, 102
video output amplifiers (TV), 129
voltage generation (TV), 130
voltage regulators, 101

W

wire sizes, 198-199
wiring harnesses, 68
shielding in, 69
shunting in, 69
woofer speakers, 36

Z

zener diodes, 47

Edited by Lisa A. Doyle